B

EXS 53:
Experientia Supplementum
Vol. 53

Birkhäuser Verlag
Basel · Boston

Development of Hormone Receptors

Edited by
György Csaba

1987

Birkhäuser Verlag
Basel · Boston

Parts of this review were published previously in 1 issue of the journal
EXPERIENTIA, Vol. 42, No. 7, 1986

Library of Congress Cataloging in Publication Data

Development of hormone receptors.
   (Experientia supplementum ; vol. 53)
   Includes index.
   1. Hormone receptors. I. Csaba, György.
II. Series: Experientia. Supplementum ; v. 53.
QP188.5.H67D48 1987   574.1'4   87-873
ISBN 3-7643-1858-9

CIP-Kurztitelaufnahme der Deutschen Bibliothek

Development of hormone receptors / ed. by György
Csaba. – Basel ; Boston ; Stuttgart : Birkhäuser,
1987.
(Experientia : Supplementum ; Vol. 53)
ISBN 3-7643-1858-9

NE: Csaba, György [Hrsg.]; Experientia / Supplementum

Teilw. in: Experientia ; Vol. 42

©1987 Birkhäuser Verlag
Printed in Switzerland
ISBN 3-7643-1858-9
ISBN 0-8176-1858-9

Contributors

Barrington, E.J.W. (formerly University of Nottingham, England)

Bückmann, D. (Universität Ulm, Federal Republic of Germany)

Carpentier, J.-L. (University of Geneva, Switzerland)

Csaba, G. (Semmelweis University of Medicine, Budapest, Hungary)

Döhler, K.D. (Bissendorf Peptide GmbH, Wedemark, Federal Republic of Germany)

Gammeltoft, S. (Université de Nice, France)

Gorden, P. (National Institute of Arthritis, Diabetes, and Digestive and Kidney Diseases, Bethesda, USA)

Gorski, J. (University of Wisconsin, Madison, USA)

Hollenberg, M.D. (University of Calgary, Canada)

Kovacs, P. (Semmelweis University of Medicine, Budapest, Hungary)

LeRoith, D. (National Institute of Arthritis, Diabetes, and Digestive and Kidney Diseases, Bethesda, USA)

Lesniak, M.A. (National Institute of Arthritis, Diabetes, and Digestive and Kidney Diseases, Bethesda, USA)

Orci, L. (University of Geneva, Switzerland)

Robert, A. (University of Geneva, Switzerland)

Roberts Jr, C. (National Institute of Arthritis, Diabetes, and Digestive and Kidney Diseases, Bethesda, USA)

Roth, J. (National Institute of Arthritis, Diabetes, and Digestive and Kidney Diseases, Bethesda, USA)

Van Obberghen, E. (Université de Nice, France)

Contents

Why do hormone receptors arise? An introduction

G. Csaba

Department of Biology, Semmelweis University of Medicine, POB 370, H-1445 Budapest (Hungary)

The hormone is a signal molecule which carries a given type of information. This information is received by a cellular signal receiver (receptor) structure, which mediates it into the cell body. Thus the information embodied by the signal (hormone) molecule acquires a 'sense', which is expressed as a cellular response. In this interpretation the hormone and its receptor form a unity, since neither of them has a 'sense' in itself. The hormones or, more precisely, the cells containing them, are the foundation stones of the hormonal system. It follows that the existence of a hormonal system presupposes the existence and, naturally, the interaction, of hormones and receptors. However, the fact that the endocrine system is an issue of evolution has prompted us to revise the concept that the hormone and its receptor could have been preexisting structures: the interaction of its corner-stones is necessarily a result of evolution itself.

The evolution of recognition

The interaction between the hormone and its receptor presupposes that they mutually recognize one another. Since, in this sense, cell-hormone recognition is a fundamental phenomenon, it seems unlikely that the receptor–hormone relationship represented the initial step of cell-environment interrelationship. Recognition processes also take place at different levels intracellularly, the simplest being the mutual recognition of the two strands of double-stranded DNA, or DNA-RNA recognition (chemical recognition in vivo). A more intricate process is enzyme-substrate recognition, in which the steric structure plays a greater role, than the amino acid sequence. A still more complex, but decisively important phenomenon is the mutual recognition between intracellular compartments. The membrane-enveloped intracellular structures seek, find, attract or repel one another, and attraction results in fusion of the membrane envelopes of compartments. Intracellular membrane fusion is a deterministic phenomenon; although its precise mechanism is still obscure, there is reason to postulate that the membrane envelopes of the intracellular compartments themselves contain certain receiver and signal molecules (markers) which account for intracellular attraction or repulsion[6].

Since DNA-DNA, DNA-RNA and enzyme-substrate recognition already appear at the lowest levels of phylogenesis (in prokaryotic cells), these phenomena may well have been involved in the origin of life, whereas compartmentalization is (apart from the single-compartment structure which satisfies the criteria of the most primitive, prokaryotic cell) the exclusive property of eukaryotic cells. Intracellular compartment formation in all probability presupposed the existence of compartment-compartment recognition, to maintain the intrinsic order. Since prokaryotes are devoid of a nuclear membrane, the source of compartment formation must necessarily have been the plasma membrane. The sugar-linked plasma membrane proteins (glycoproteins) may have been responsible from the very beginning for cell-environment recognition[14, 31], and for intracellular recognition as well. The recognition capacity of the cytoplasmic membrane is also dual in present-day living beings. Receptor-mediated endocytosis (internalization) accounts for transport of the receptor-bound structures into the cell inside coated vesicles, which themselves contain structures capable of recognition, or of being recognized, thus determining the fate of the material transported inside them. It appears that, although intracellular (compartment-compartment) recognition *virtually* represents a lower level of phylogenesis than cell-environment recognition, it has in fact developed from the latter.

Recognition of the environment is decisive for the cell at all levels of phylogenesis. All life-conditions of the cell are furnished by its environment, which simultaneously serves as the scource of nutriment and as the sink into which the cellular degradation products are released, and contains both useful and noxious materials. From the evolutionary point of view, that cell which fails to distinguish between the advantageous and noxious qualities of the environmental molecules is doomed to deterioration and dies without producing any progeny[7]. *Only those cells which can fully adapt themselves to their environment are capable of multiplication. Cell-environment recognition is therefore a fundamental prerequisite of evolution.*

Signal receivers and signal molecules

It is known from endocrinology that cells possess certain well-defined receptor structures which are capable of interaction with given materials (hormones). The question arises whether such predetermined interaction was possible in the initial stage of evolution, which took place primarily in an aquatic environment. The answer is unequivocally no, because free water always contains a practically infinite variety of dissolved materials. The cells representing the lowest levels of phylogenesis, regardless of whether they are prokaryotic or eukaryotic, are capable of moving from one place to another which involves a change in the quality of environmental materials (signal molecules) as well. In this light the cell-membrane-associated signal receivers (receptors) cannot be interpreted as preformed, stable structures, but rather as transient patterns arising by the continuous dynamic change of the cell membrane, and 'questioning' the (given) environment. The fluid mosaic membrane, characteristically present already at the unicellular level, makes possible not only the movement of membrane proteins, but also their assembly in different configurations. In this light the dynamic receptor theory of Koch and co-workers[21], according to which different membrane patterns capable of acting as signal receivers can arise by assembly of sub-patterns in the dynamically changing fluid mosaic membrane, seems to explain the mechanism of receptor formation.

This interpretation does not exclude the hypothesis that, on the other hand, environmental materials (signal molecules) are also capable of recognizing a complementary membrane pattern, and lead to its amplification if the interaction between them – mediation to the cell of the signal represented by the environmental molecule – can take place under the given conditions[2, 3, 5]. The intracellular (cytosolic) post-receptor mediation system operates at all levels of phylogenesis, to mention only cyclic AMP, or the Ca^2-calmodulin system[10, 19, 20, 22, 23, 29, 33]. Receptor formation and/or amplification in presence of the specific signal molecule has been demonstrated experimentally in unicellular model systems, and evidence has also been obtained that the selection advantage of the cells presenting the adequate receptor pattern promotes the establishment, and evolutionary amplification, of the given receptor-hormone relationship[12]. The unicellular Tetrahymena responds to primary membrane-level interaction with a foreign molecule by increased division (higher mitotic rate), which persists over many subsequent generations. The evolutionary significance of this phenomenon is the transmission of the newly acquired information to an increased number of offspring. Thus the acquired information (hormonal or signal imprinting) becomes fixed, and the second interaction with the given signal molecule accounts for a further increase in the given cellular function over as many as 500 generations[13].

At least as important as dynamic receptor formation at the unicellular level is the presence of genetically encoded stable hormone receptors, and of a pre-programmed receptor-hormone relationship, in multicellular organisms. The conditions of signal reception serve at both levels as tools of adaptation and survival of the species (individual), which carries the receiver system. One is, therefore, obliged to postulate that, from the evolutionary point of view, the stimulus for receptor formation is the presence of the signal molecule (the future hormone), and that, once established, receptor structures persist and become stabilized if their existence, and the receptor-hormone relationship which they make possible, is advantageous for the cell, or for the multicellular organism of which the cell forms part. This does not, however, exclude that once the receptor is encoded it can, at a given stage of its development, also require the presence of the hormone. This will be explained later[8].

Interrelation of hormone and hormone receptor evolution

One might conclude from the foregoing considerations that any molecule could develop into a hormone, and any hormone (signal molecule) could induce the formation of a specific receptor. However, in reality, the hormone families are relatively small in number, and relatively few materials are utilized as signal molecules[1]. It follows that certain steric structures are obviously privileged in respect of forming cellular receptors for themselves and entering into a receptor-hormone relationship. Moreover, certain molecules which have steric structures which would make them well able to act as signal molecules, are reserved for other (non-signal) functions in the living organisms. Thus it appears that only a given set of molecules is capable of the signal function. For example, certain molecules not belonging to the neuroendocrine system are also bound at the receptor level and can even evoke a cell-mediated response, but are excluded from the receptor-hormone relationship category for reasons of system theory (receptor-mediated endocytosis, transferrin, LDL receptors, etc.). Although the foregoing considerations seem to suggest the priority of the signal molecule against the signal receiver structure, on the grounds that the presence of the former stimulates the formation of the latter, it should be noted that of the two factors the once-established receptor seems to be the more stable structure, because changes in the quality of the signal molecules have been regularly demonstrated in the course of evolution[18, 27, 28]. Thus in present-day higher organisms the receptors bear a close resemblance to their primordial predecessors, whereas the signal molecules seem to have acquired their present role in a long evolutionary process. At the same time, the present-day signal molecules are able to bind to primitive receptor structures, too. This signifies not only a certain plasticity of the receptor's binding capacity, but also a lesser alteration of hormone quality than the alteration that would be needed to exclude immunological cross-reactions between present-day and primordial configurations.

Consideration should also be given to the fact that the current classification of the signal molecules (hormones) is largely anthropocentric, or rather vertebrate-centric, inasmuch as exclusively the signal molecules existing (identified) as such in higher organisms are regarded as hormones, whereas their less-developed forms, which also occur in vertebrates, are regarded as precursors. The precursor conception may, however, often be misleading, because at lower levels of phylogenesis the so-called precursor molecule may represent a much stronger signal (for cells of that level) than the (vertebrate) hormone proper, since both parts of the receptor-hormone system are closer to the initial stage of evolution[9, 11]. Thus the classification of certain molecules as hormone or hormone precursor frequently applies only at a given level of phylogenesis (or to a given organism) and expresses a given stage in the development of the system. While the signal molecule's role and importance in receptor formation is practically an established fact, the factors accounting for the hereditary transmission of the receptor are still obscure. The unicellular organism can, with the help of its membrane, recognize and 'memorize' the foreign molecule, which thereby acquires the role of a signal molecule, but explanations of the mode of transmission of this 'memory' to the progeny generation remain hypothetical[5, 7]. If the information is membrane-associated, it should dissipate in the course of serial divisions, but this is definitely not the case. Probably the membrane-associated information becomes fixed via self-assembly of the membrane proteins. This explanation seems feasible, but fails to account for the ontogenetic encoding of the receptor, unless a gene-level fixing of some form of the membrane-associated information is postulated. However, this cannot be reconciled with the current genetic conceptions, for it implies the hereditary transmission of an acquired property. It is expected that advances in research into immunological memory[16, 24] and membrane DNA[17, 32] will also throw more light on the receptor 'memory' problem.

At all events, in the light of the new interpretation of the hormone and precursor categories, the gene-level fixing of receptor 'memory' seems to be the prerequisite of those mutations which account for the existing differences between the present-day and primordial receptor structures, and thereby for the greater binding affinitiy shown by mammalian receptors to the mammalian hormones than to the 'precursors' of these.

After binding the ligand, the membrane receptors become internalized, and after internalization, they are recycled to the membrane again[34]. This mechanism makes possible the degradation of the ligand, and thereby the termination of its action. Thus internalization of the membrane receptor seems to be the ontogenetic recurrence of the phylogenetic internalization process[4], which ultimately led to the formation of intracellular membrane structures (compartments). This hypothesis leads to further speculations, for, according to present knowledge, hormone receptors occur in two locations: membrane-associated, and cytosolic. Recent experimental observations have increasingly suggested, in contrast to earlier conceptions, that any hormone is capable of binding to both types of receptors, e.g. steroids can bind to membrane receptors and polypeptide hormones to cytosolic receptors[15,25,26,30,35,36]. Since, logically, all materials interacting with the cell come into contact first and foremost with its membrane, it might well be postulated that at the beginning of evolution exclusively membrane receptors existed, which gained access into the body of the cell only at a later stage, probably in a dissolved form, and gave rise therein to cytosolic receptors. Thus, although evidence is lacking, theoretically the cytosolic receptors can be regarded as descendants of the membrane receptors, which arose from the necessity of refining the mechanisms of hormone binding and hormone transport[4].

Conclusions

In view of the foregoing considerations, it appears that the receptor-hormone relationship is, by origin, essentially a cell-environment (chemical) relationship which influences cell behavior. With the development of multicellularity, the interests of the single (individual) cell became subordinated to those of the cell population (community), and the cell-environment relationship became modified inasmuch as receptor activity became integrated into the functional program of the entire organism. Accordingly, the 'open program' of the individual cell, which involved continuous dynamic changes of the membrane receptors under the influence of the signal molecules, was superseded by a 'closed program' for the given receptor, which gave rise to a chemical memory of the cell. With multicellularity the cellular functions have become integrated into an almost entirely predetermined program in which the quality and operation of the receptors are encoded to maintain the system of regulation, and impart differentiating features to given types of target cells which distinguish them from others, and delimit the response potentials of the species. A limited openness of the pre-programed system exists in the early stage of ontogenesis, and accounts for certain individual variations within the limited potentials of the species.

The answer to the question posed in the title of this paper is therefore the following: the hormone receptors arise because the external environment of the individual cell is transformed at the multicellular level to an internal environ-

12

ment, in which the random variety of environmental molecules is replaced by a predetermined set of ligands (signal molecules). Under these conditions the randomly-presented membrane patterns capable of signal reception are transformed to encoded receptor structures which execute a programed function of the closed system, but nevertheless preserve some primordial traits, which can explain many surprising observations in the field of receptor physiology.

1 Barrington, E. J. W., Evolutionary aspects of hormone structure and functions, in: Comparative Endocrinology, pp. 381–396. Eds J. P. Gaillard and H. Boer. Elsevier, North Holland Amsterdam 1978.
2 Csaba, G., Phylogeny and ontogeny of hormone receptors: the selection theory of receptor formation and hormonal imprinting. Biol. Rev. 55 (1980) 47–63.
3 Csaba, G., Ontogeny and phylogeny of hormone receptors. Karger, Basel/New York 1981.
4 Csaba, G. Newer theoretical considerations of the phylo- and ontogenetic development of hormone receptors. Acta biol. hung. 31 (1980) 465–474.
5 Csaba, G., The present state in the phylogeny and ontogeny of hormone receptors. Horm. Metab. Res. 16 (1984) 329–335.
6 Csaba, G., The development of recognitions systems in the living world. Karger Gazette 46–47 (1984) 14–16.
7 Csaba, G., The unicellular Tetrahymena as model cell for receptor research, Int. Rev. Cytol. 95 (1985) 327–377.
8 Csaba, G., Receptor ontogeny and hormonal imprinting, Experientia 42 (1986) 750–759.
9 Csaba, G., Bierbauer, J., and Fehér, Z., Effect of melatonin and its precursors on the melanocytes of planaria (Dugesia lugubris) Comp. Biochem. Physiol. 67C (1980) 207–209.
10 Csaba, G., and Nagy, S. U., Effect of vertebrate hormones on the cyclic AMP level in Tetrahymena. Acta biol. med. germ. 35 (1976) 1399–1401.
11 Csaba, G., and Németh, G., Effect of hormones and their precursors on protozoa – the selective responsiveness of Tetrahymena. Comp. Biochem. Physiol. 65B (1980) 387–390.
12 Csaba, G., Németh, G., Juvancz, I., and Vargha, P., Involvement of selection and amplification mechanisms in hormone receptor development in a unicellular model system. BioSyst. 15 (1982) 59–63.
13 Csaba, G., Németh, G., and Vargha, P., Development and persistence of receptor 'memory' in a unicellular model system. Expl Cell. Biol. 50 (1982) 291–294.
14 Damsky, C. H., Knudsen, K. A., and Clayton, H. B., Integral membrane glycoproteins in cell-cell and cell-substance adhesion, in: The biology of glycoproteins, pp. 1–64. Ed. R. J. Ivatt. Plenum Press, New York/London 1984.
15 Diez, A., Sancho, M. J., Egana, M., Trueba, M., Marino, A., and Macarella, J. M., An interaction of testosterone with cell membranes. Horm. Metab. Res. 16 (1984) 475–477.
16 Fristrom, J. W., and Spieth, Ph., in: Principles of Genetics, pp. 500–502. Blackwell, Oxford 1980.
17 Gabor, G., and Bennett, R. M., Biotin labelled DNA: a novel approach for the recognition of a DNA binding site on cell membranes. Biochem. biophys. Res. Commun. 122 (1984) 1034–1039.
18 Ginsberg, B. H., Kahn, C. R., and Roth, J., The insulin receptor of the turkey erythrocyte: similarity to mammalian insulin receptors. Endocrinology 100 (1977) 520–525.
19 Karyia, K., Saito, K., and Iwata, H., Adrenergic mechanism in Tetrahymena III. cAMP and cell proliferation. Jap. J. Pharmac. 24 (1974) 129–134.
20 Kassis, S., and Kindler, S. H., Dispersion of epinephrine sensitive and insensitive adenylate cyclase from the ciliate Tetrahymena pyriformis. Biochim. biophys. Acta 391 (1975) 513.
21 Koch, A. S., Fehér, J., and Lukovics, I., Single model of dynamic receptor pattern generation. Biol. Cybernet. 32 (1979) 125–138.
22 Kudo, S., and Nozawa, Y., Cyclic adenosine 3',5'-monophosphate binding protein in Tetrahymena: properties and subcellular distribution. J. Protozool. 39 (1983) 30–36.
23 Kuno, T., Yoshida, C., Tonaka, R., Kasai, K., and Nozawa, Y., Immunocytochemical localization of cyclic AMP in Tetrahymena. Experientia 37 (1981) 411–413.
24 Leder, P., The genetics of antibody diversity. Scient. Am. 246 (1982) 72–83.
25 Levey, G. S., and Robinson, A. G., Introduction to the general principles of hormone-receptor interaction. Metabolism 31 (1982) 639–645.

26 McKerns, K. W., Regulation of gene expression in the nucleus by gonadotropins, in: Structure and function of the gonadotropins, pp. 310–338. Ed. K. W. McKerns. Plenum Press, New York 1978.

27 Muggeo, M., Ginsberg, B. H., Roth, J., Neville, G. M., Meyts, P. de, and Kahn, C. R., The insulin receptor in vertebrates is functionally more conserved during evolution than the insulin itself. Endocrinology *104* (1979) 1313–1402.

28 Muggeo, M., Obberghen, E. van, Kahn, C. R., Roth, J., Ginsberg, B., Meyts, P. de, Emdin, S. O., and Falkmer, S., The insulin receptor and insulin of the atlantic hagfish. Diabetes *28* (1979) 175–181.

29 Nagao, S., Suzuki, Y., Watanabe, Y., and Nozawa, Y., Activation by a calcium-binding protein of guanylate-cyclase in Tetrahymena pyriformis. Biochem. biophys. Res. Commun. *90* (1979) 261–268.

30 Rao, C. V., and Chegini, N., Nuclear receptors for gonadotropins and prostaglandins, in: Evolution of hormone receptor systems, pp. 413–423. Alan. R. Liss, New York 1983.

31 Reading, C. L., Carbohydrate structure, biological recognition and immune function, in: The biology of glycoproteins, pp. 235–321. Ed. R. J. Ivatt. Plenum Press, New York/London 1984.

32 Reid, B. L., and Charlson, A. J., Cytoplasmic and cell surface deoxyribonucleic acid with consideration of their origin. Int. Rev. Cytol. *60* (1979) 27–52.

33 Satir, B. H., Garofalo, R. S., Gillingan, D. M., and Maihle, N. J., Possible functions of calmodulin in protozoa. Ann. N.Y. Acad. Sci. USA *356* (1980) 83–93.

34 Steinmann, R. M., Melmann, I. S., Muller, W. A., and Cohn, A., Endocytosis and the recycling of plasma membrane. J. Cell Biol. *96* (1983) 1–27.

35 Szego, C. M., Parallels in the modes of action of peptide and steroid hormones: membrane effects and cellular entry, in: Structure and function of the gonadotropins. pp. 471-472. Ed. R. W. Mc Kerns. Plenum Press, New York/London 1978.

36 Szego, C. M., and Pietras, R. J., Lysosome function in cellular activation: propagation of the actions of hormones and other effectors. Int. Rev. Cytol. *88* (1984) 1–302.

Mechanisms of receptor-mediated transmembrane signalling

M. D. Hollenberg

Endocrine Research Group, Department of Pharmacology and Therapeutics, University of Calgary, Faculty of Medicine, 3330 Hospital Drive N.W., Calgary (Alberta T2N 4N1, Canada)

Receptors, acceptors, channels and the problem of transmembrane signalling

Fundamental to the successful function of any multicellular organism is an efficient communication system that can convey information from one cell to another. Although the overall function of the cell membrane is to maintain an effective barrier between the intracellular and extracellular milieu, highly specialized membrane structures (e.g. ion channels, nutrient transporters, histocompatability determinants) can be singled out as playing particularly pivotal roles in terms of selectively transmitting information from the external to the internal cell environment (and in some cases, vice versa). Over the past decade there has been much progress in the biochemical and pharmacologic characterization of the membrane constituents that participate in the transmembrane signalling process. This chapter will deal in general with selected aspects of transmembrane signalling and will focus in particular on the plasma membrane-localized processes used by pharmacologic receptors.

In some cases, information is transferred simply by the selective uptake of specific ions or nutrients via transmembrane channels. The energy for this transfer process may come either from a chemical concentration gradient of the transported ligand or from a coupled process in which cellular energy (e.g. in the form of ATP) is used to 'drive' the ion or metabolite across the membrane. In such cases, the ion or metabolite *is* the message that is being communicated to the intracellular space, and it is up to the intracellular milieu to respond to increased (or decreased) concentrations of the transported substance and to regulate the rate (or amount) of transport that may occur. It is now evident that the membrane constituents that act as selective transport molecules are highly complex proteins that may be subject to the same kind of allosteric regulation as are cellular enzymes. The biochemical and pharmacologic characterization of the voltage-sensitive sodium channel, with its complex toxin binding sites[7, 34a], may serve as a useful model for the characterization of channels for other ions and metabolites.

In a number of cases, substances that convey information for the regulation of intracellular processes (for example, vitamin B_{12} or cholesterol) are associated with carrier proteins in the blood stream (e.g. transcobalamin-II for B_{12}; low density lipoprotein, or LDL for cholesterol). The blood-borne carrier proteins can be recognized in a very specific high-affinity manner by specialized transporters responsible for the selective adsorptive pinocytosis of the regulatory ligands (the examples being used here are B_{12} and cholesterol). Here again, as in the case of the sodium ion, the ligand transported (i.e. cholesterol or B_{12}) *is* the message being taken into the cell. The binding constituents (either the carrier proteins or the cell-surface uptake proteins) perform a message-carrying function and do not, of themselves, generate a transmembrane signal. Thus, the binding constituents for substances like the LDL-cholesterol complex[3, 4] or the transcobalamin-II-B_{12} complex[32, 48] while commonly referred to as 'receptors', do not truly function as receptors in the pharmacologic sense. It has proved convenient to refer to such highly specific transporters either as 'acceptors'[17] or receptors of the class II type[25].

In contrast to acceptors, pharmacologic receptors located in the plasma membrane exhibit not only the ability to recognize a ligand with high affinity and selectivity, but also the capacity, once combined with the specific ligand, to participate in the process of cell activation. It is this dual recognition-activation property, by which the receptor per se acts in part as a message-generating system, that distinguishes receptors from other cell surface recognition/transport constituents. The remainder of this article will focus on the general mechanisms whereby cell surface receptors generate a transmembrane signal. Receptors for steroid hormones will not be discussed.

General mechanisms of transmembrane signalling

1. Basic mechanisms. In order to fulfill their recognition/activation function, membrane receptors must be able to generate an intracellular signal that can be greatly amplified. Based on observations made over the past decade, it is possible to generalize somewhat concerning the mechanisms employed by receptors for transmembrane signalling. Structural data obtained to date indicate that those constituents capable of generating a transmembrane signal actually have transmembrane domains that can either communicate with (e.g. via an ion channel) or interact with the intracellular/submembraneous environment. Thus, a physical contiguity between an external and internal receptor domain may turn out to be a prerequisite for a transmembrane signalling molecule. Constituents that exist solely on the external aspect of the plasma membrane may be able to generate intracellular signals only via an interaction with a second membrane-localized component that possesses a transmembrane domain.

The mechanisms whereby the transmembrane receptor domain generates the intracellular signal may turn out to be few in number. Based on data accumulated to date, one can single out three basic processes, as illustrated in figure 1: a) ligand-modulated ion channel activity, b) ligand-regulated enzymatic activity, with the catalytic domain situated in the intracellular portion of the receptor and c) ligand-regulated liberation of cryptic mediators via interactions of intracellular receptor domains with other submembraneous constituents like the so-called G- or N-regulatory components of the adenylate cyclase system (see below). The process of receptor internalization, whereby the receptor-bound ligand finds its way into intracellular organelles[36, 37], provides an intriguing mechanism whereby one of the three basic reactions performed by receptors can

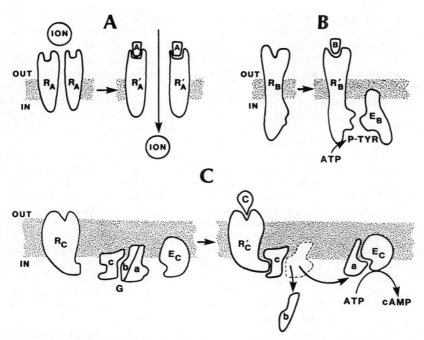

Figure 1. Hypothetical models of ligand-activated transmembrane signalling using three basic mechanisms. *A* Ligand-regulated ion channel, like the nicotinic cholinergic receptor. *B* Ligand-regulated enzyme activity. Receptor tyrosine kinase is used as the example, wherein phosphorylation of a membrane effector, E_B plays a role in signalling. *C* Ligand-modulated release of cryptic mediators. Upon interaction with a ligand-occupied receptor, a hypothetical guanine nucleotide regulatory complex (G) is shown to release an effector-stimulatory component (a) and at least one other regulatory protein (b). In this instance, adenylate cyclase is depicted as the effector (E_C). In principle, any other membrane process (ion channel, phosphodiesterase, phospholipase) could be similarly regulated. Models are shown for receptors in either the absence (R_A, R_B, R_C) or presence (R_A', R_B', R_C') of their regulatory ligands (A, B, C). None of the components are drawn to scale. The extracellular (out) domains of the receptors face the top of the page; the intracellular (in) receptor domains, anchored in the membrane (stippled areas) face the bottom of the page.

be translocated to a targeted intracellular environment. A key element of receptor function, irrespective of which of the above cited mechanism is used, is the basic requirement of the ligand (or its surrogate) to act as an allosteric regulator of the receptor mechanism.

2. Signal amplification. A fundamental question to answer is: how does the cell amplify the signal generated by each of the three basic processes outlined above? Fortunately, information acquired to date provides a number of answers to this question for each basic mechanism. For instance, a change in membrane potential, resulting from a ligand-induced change in ion flux (panel A, fig. 1) would, as elaborated upon by Zierler and colleagues[56, 57], have a profound effect on the orientation of many membrane proteins, thereby rapidly generating an overall intracellular signal. The voltage-sensitive sodium channel[7, 34a] provides an interesting example of the effect of membrane potential on membrane protein function, with consequent signal amplication. Alternatively, a ligand-regulated channel for an enzyme-regulatory cation like calcium could rapidly control intracellular calcium-modulated enzymes that participate in cascade reactions vital to cell regulation. In principle, ligand-modulated channels for either anions or cations could lead to such amplified intracellular signals. The distinction between the receptor channel activity and the activities of other ion/metabolite channels rests entirely on the ligand-regulated properties of the receptor channels. The nicotinic cholinergic receptor represents the best understood ligand-regulated channel of this kind[9, 30, 31, 34, 52].

It is comparatively easy to visualize how the second basic receptor process, ligand-regulated enzymatic activity could yield an amplified receptor signal. The discovery that receptors for epidermal growth factor-urogastrone (EGF-URO)[5, 6], insulin[26, 40, 41, 55] and platelet-derived growth factor (PDGF) are ligand-regulated tyrosine kinases (for a brief review, see ref. 19), indicates that phosphorylation/dephosphorylation cascade reactions, known to play a vital role in many metabolic pathways, probably also play a central role in receptor function. Such cascade reactions, initiated by ligand-triggered tyrosine kinase activity (panel B, fig. 1) could readily amplify the initial receptor stimulus. Apart from kinase activities, any other enzymatic activity (e.g. proteolytic activity; phospholipase activity) could, in principle, be used by a receptor system to initiate a cascade signal amplification process. To date, there is not yet evidence for or against the existence of receptors with such intrinsic enzymatic activities. Such ligand driven enzymatic processes could readily generate compounds (e.g. peptides; diacylglycerol; archidonic acid) that might serve as intracellular 'second messengers'. It will be of great interest in the future to look for receptors with intrinsic enzymatic activities other than that of the kinase class.

Those receptors that regulate the production of the second messenger, cyclic AMP, represent well-known examples of receptors that utilize the third basic process outlined above. Only recently has it been appreciated that the catalytic activity of the adenylate cyclase system is not controlled directly by the receptor itself, but rather by an indirect process, whereby the receptor liberates a separate cyclase-regulatory polypeptide (so-called α_s[15, 35]). The liberation of such cryptic regulatory polypeptides from the so-called guanine nucleotide binding regulatory proteins (G-proteins or N-proteins[15, 39]) can provide for the concerted regulation of a variety of cell activities apart from adenylate cyclase. A generalized case for this mechanism is illustrated in panel C of figure 1. The oligomeric structure of the G-proteins (known to contain α, β and γ subunits) and the discovery of several distinct -substituents[35] provides for enormous flexibility in the use of such a transmembrane signalling process. In terms of the cyclic AMP system, the amplification reactions involving cyclic AMP-dependent protein kinases have been widely studied. This represents perhaps the best-understood (biochemically) receptor-mediated signalling system. However, because the common β-substituent can combine with (and presumably regulate) a variety of the family of α-substituents, and because the metabolic functions of the α-subunits other than the cyclase-regulatory α_s-substituent have yet to be determined, it appears likely that the cyclase system may represent only the 'tip of the iceberg' in terms of the regulation of cell function by this third distinct basic receptor-signalling process. One looks forward eagerly to interesting developments in this area of study.

3. Messengers. As pointed out in an earlier section, in the situation where information transfer is mediated via ion/metabolite channels, the ion itself represents 'the message'. In contrast, receptor-mediated information transfer involves the generation of at least one, and perhaps multiple diffusible messengers. In one sense, for the action of hormones like insulin, the receptor per se can be called the 'second messenger' (the hormone itself may be thought of as the 'first messenger' in the communication chain leading from hormone binding to cellular activation). For some time, cyclic AMP was designated as a 'second messenger' for the action of hormones like glucagon or epinephrine. Now, however, it can be appreciated that the so-called α_s-stimulatory subunit of the cyclase-regulatory guanine nucleotide binding complex[15, 35] also participates as a messenger in the course of the action of either glucagon or epinephrine. Further, it is not unlikely, as elaborated upon in a subsequent section, that a single hormone-receptor combination may initiate not one, but perhaps many membrane-localized reactions. Thus, it may be inappropriate to single out any particular diffusable low-molecular-weight compound as a primary messenger for any hormone; rather, it may be more fruitful to think in terms of a characteristic matrix of diffusable messengers generated by the combination of a single hormone with its receptor. Compounds that have been singled out for particular attention in terms of transmembrane signalling processes are: 1) Sodium ion (nicotinic cholinergic receptor); 2) Cyclic AMP (cyclase-related hormones); 3) Cyclic GMP (several neurotransmitter receptors; including the muscarinic cholinergic receptor); 4) Calcium (for neurotransmitter receptors, including muscarinic cholinergic as well as for other peptide receptors like those for angiotensin and pancreozymin); 5) diacylglycerol; 6) inositol trisphosphate; 7) α-substituents of the G-protein complex[15, 35]; and 8) as yet unidentified peptide regulators related to the action of insulin[23, 28, 43, 47]. It is important to note that in the case of many neurotransmitters and hormones, not one, but three of the above identified 'messengers', namely calcium, diacyglycerol and inositol trisposphate may be involved in a complex bifurcating signal pathway, involving the hydrolysis of membrane phosphoinositides[1]. A critical factor in understanding the chain of information transfer via the receptor to the cell relates to the identification of the key targets of the messengers, viz. the protein kinases regulated by cyclic AMP; kinase C regulated by diacylglycerol[33]; intracellular calcium-binding proteins from which calcium is released in the presence of inositiol triphosphate[1]; and calcium-modulated regulatory enzymes or proteins like calmodulin that may have a widespread effect on overall cell regulation. Possibly, only a handful of such messengers, generated via membrane-localized reactions will be involved in the action of most neurotransmitters and hormones.

Receptor dynamics and transmembrane signalling

1. Receptor mobility and transmembrane signalling. In the not-too-distant past, possibly due to the focus of pharmacological studies on nerve-muscle preparations, receptors were thought of as specifically localized entities (e.g. at the neuromuscular junction) that were tacitly assumed to be in a more or less static state as a consequence of cell differentiation. Now, however, it is realized that in perhaps the majority of cases, receptors for agents such as insulin and epidermal growth factor-urogastrone (EGF-URO) are dynamic cell surface constituents that can migrate in the plane of the cell membrane. Both pharmacologic receptors and acceptors (like the one for LDL-cholesterol) appear to share this mobile property. In studies with cultured cells (e.g. mouse fibroblasts) it has been observed that subsequent to the binding of a ligand-like insulin, a receptor (or acceptor) can undergo a complex series of protein-protein interactions leading to the cellular internalization and intracellular processing (e.g. degradation; or possibly recycling to the cell surface) of both the receptor and the bound ligand[27, 36, 37]. The concept of a hormone receptor as a 'mobile' or 'floating' membrane constituent evolved along with the development of understanding of the general properties of cell surface proteins. In terms of hormone-triggered transmembrane signalling, receptor mobility is viewed as a most important property that can enable the receptor to interact with a variety of membrane constituents in the course of cell activation. It is a basic tenet of the 'mobile' or 'floating' receptor paradigm of hormone action[2, 10–12, 17, 22] that the binding of a ligand dramatically alters the ability of a receptor to migrate in the plane of the membrane and to interact with other membrane components. Thus, the entity, [HR], resulting from the combination of a hormone with its receptor:

$$[H]+[R] \rightleftharpoons [HR]$$

can go on to form ternary complexes, of the kind, [HRE]:

$$[HR]+[E] \rightleftharpoons [HRE]$$

wherein E represents an effector molecule involved in the process of cell activation. The hormone-receptor complex may also undergo an isomerization reaction ($[HR] \rightleftharpoons [HR^*]$) as has been suggested for the muscarinic receptor[50a]. A number of variations of this model have been developed[22]. For instance, the above equations illustrate an 'association' model, wherein the formation of hormone-receptor complex promotes receptor-effector coupling. An alternative possibility is a 'dissociation' model, wherein an inactive effector, held in a receptor-effector complex, RE is dissociated when the ligand binds, to yield a free active effector, E^*:

$$H+RE \rightleftharpoons HR+E^*$$

In principle, the mobile receptor model does not restrict the number of distinct effector moieties with which the ligand receptor complex might interact. In this manner, by interacting with multiple effectors, a single hormone receptor complex could liberate simultaneously a variety of intracellular mediators.

2. Receptor microclustering, patching and internalization. Largely stimulated by the 'floating' or 'mobile' receptor model, summarized in the previous section and heralded by the work of Edidin and collaborators[14], much work has focussed on measurements of receptor mobility. With the use of fluorescent ligand probes (insulin, EGF-URO) lateral diffusion coefficients in the range of 5×10^{-10} cm^2/s have been observed[27, 36, 44]. Studies using radiolabeled ligands or ligands tagged with heavy metals (gold; ferritin) have also visualized either preclustered or ligand-induced receptor clusters in the cell membrane. The speed of receptor diffusion is sufficient to allow the ligand-occupied receptor to collide with many other membrane components in a very short time period (e.g. tens of milliseconds).

From observations with a number of ligands, it appears that many receptors follow a common sequence of mobile reactions subsequent to the ligand binding event (fig. 2). In the absence of their specific ligands, receptors can be diffusely distributed over the cell surface. However, as illustrated in figure 2, at physiological temperatures, the binding of a ligand can lead to a rapid microclustering (receptor microclusters, containing perhaps 2 to 10 receptors) and a reduction in receptor mobility, accompanied by the progressive aggregation of ligand receptor complexes into immobile patches (aggregates containing 10's to 100's of receptors) that can be visualized by fluorescence photomicrography. In cultured fibroblasts, the microclustering event is thought to preceed the formation of patches that can be seen in the fluorescence microscope. In terms of the process of transmembrane signalling, to be discussed below, the microclustering event (formation of n-mers containing groups of 2 to 10 receptors) is viewed as a phenomenon related to, but separate from the aggregation process (formation of patches, containing 10's to 100's of receptors). Subsequent to the formation of the comparatively large receptor aggregates, the ligand-receptor complexes can be either shed into the medium or taken into the cell (internalized). Receptor internalization appears to be an ongoing process that is accelerated when a ligand such as insulin binds to its receptor. It is not clear whether or not receptor occupation is a prerequisite for forming small receptor clusters in all cell types. For instance in adipocytes[50], there are data to indicate that insulin receptors exist as clusters prior to the addition of insulin. The mechanism(s) that lead to microclustering, aggregation and internalization of receptors (or acceptors) are poorly understood. In many cells, such as fibroblasts, internalization appears to occur at specific sites on the cell surface – the so-called 'bristle-coated pit'[36, 37]. In some cell types (e.g. adipocytes) receptors may be localized and internalized at sites other than the 'coated pit' regions[50].

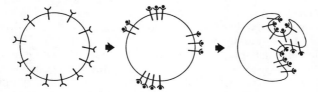

Figure 2. Microclustering and patching of receptors: early events related to cell activation. The scheme illustrates the ligand-mediated microclustering and patch formation processes described in the text. Upon going from the predominantly diffuse distribution (hormone-free state) to the ligand-occupied microclustered state, it is thought that in many cells, such as fibroblasts, ligand-receptor complexes coalesce in coated pit regions prior to cellular internalization. It is important to note that in some cell types receptors may exist in a preclustered state and that in certain cells, receptor internalization can occur predominantly at sites distinct from the coated pit region[50].

Subsequent to aggregation, the receptor can be internalized via an endocytic process into a cellular compartment that appears to be distinct from the lysosome (fig. 3). The intracellular receptor-bearing vesicles, which in contrast with

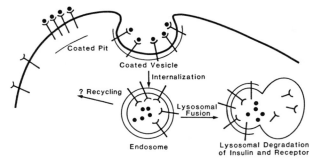

Figure 3. Formation and migration of receptor-bearing vesicles. The aggregated receptors are thought to become trapped in an endocytic vesicle that buds inward, so as to form an intracellular vesicle, the endosome (or receptosome). Evidence suggests that the internal environment of the endosome becomes acidified, favoring the dissociation of many ligands, e.g., insulin, from their receptors. After the initial internalization event, the receptor-bearing vesicles are thought to change their shape (possibly by fusing with other nonlysosomal intracellular constituents) and to migrate to a variety of cellular locations. Two further possibilities are also depicted: fusion with lysosomal structures and recycling of the receptor to the cell surface.

lysosomes are not phase-dense in the electron microscope and are acid phosphatase negative, have been termed 'endosomes' or 'receptosomes'[36]; the latter term emphasizes the role of these specialized endocytic vesicles in the process of receptor-mediated endocytosis. One possible fate of such receptor-bearing endosomes is fusion with lysosomes, followed by the lysosomal degradation of the receptor (so-called receptor processing) and of the bound ligand. Several studies suggest that limited receptor processing may also occur at a prelysosomal site (possibly, endosome-associated). An alternative route that the receptosome may follow leads back to the cell surface via a recycling process that reintegrates the receptor into the plasma membrane. A possible fusion of the receptosome with other intracellular organelles (e.g. nuclear envelope) cannot be ruled out, but has yet to be documented. At present, little is known about the factors that control either the internalization process or the trafficking process that may lead on the one hand to lysosomal receptor processing or, on the other hand, to a recycling of the receptor back to the cell surface (fig. 3). The intracellular receptor domains may play an important role in this trafficking process. There is also little known about the possible role(s) for the degradation products (ligand or receptor fragments) that may be released into the cytoplasma as a result of the endosomal and lysosomal degradation (processing) events. In view of the paripatetic nature of the hormone-receptor complex, migrating from the cell surface to the cytoplasmic space, a key question to answer is: What role (if any) do these receptor-migratory pathways play in the process of transmembrane signalling? The following sections will deal with this question.

Receptor microclustering and cell activation

It has been appreciated for some time that ligand bivalency (or multivalency) is critical for the patching and capping of surface macromolecules[38]. Only comparatively recently, however, has it become apparent that pharmacologic receptors can patch and cap, and that crosslinking (or microclustering) may be a key event in the process of transmembrane signalling. Some of the most impressive data implicating receptor crosslinking as a key event have come from experiments using polyclonal anti-insulin receptor antibodies (derived either from insulin-resistant patients[24]; or from rabbits immunized with purified insulin receptor[27]).

In brief, the results with the polyclonal anti-receptor antibodies are summarized in figure 4. Both intact antibody (IgG) and the bivalent (Fab)$_2$ derivative were capable of stimulating cells. However, the monovalent Fab derivative was not biologically active even though it acted as a competitive inhibitor of insulin binding[24]. Strikingly, the biological activity of the monovalent Fab fragment was restored by the addition of a second bivalent antibody directed against the Fab fragment (fig. 4). In essence, the crosslinking of the receptor-associated antibody Fab binding domain appears to be required for cell activation. The anti-insulin receptor antibodies are capable of mimicking many of the actions of insulin in target cells[24]. Thus, the work with the antireceptor antibodies led to two conclusions: 1) the receptor alone and not the ligand (e.g. insulin) possesses the information required for transmembrane signalling and 2) receptor microaggregation appears to be a key event related to transmembrane signalling. As mentioned above, the mechanism(s) whereby the binding of an agent like insulin leads to receptor crosslinking are as yet poorly understood.

Evidence implicating receptor microclustering as an important event for cell activation has now been obtained for other peptide hormones (EGF-URO and luteinizing hormone-releasing-hormone: LHRH or GnRH). In one series of experiments, it was observed that a chemically modified derivative of epidermal growth factor-urogastrone (CNBr-EGF-URO, see below), that of itself bound

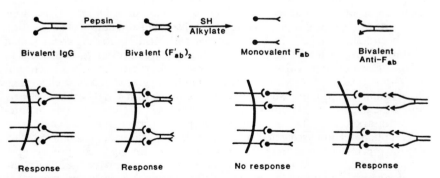

Figure 4. Effects of divalent and monovalent antireceptor antibodies. These results were obtained with polyclonal anti-insulin receptor antibodies[24]. Intact antibody (IgG) and the bivalent antibody derivative generated by pepsin cleavage (F'$_{ab}$)$_2$ were capable of stimulating cells. However, the monovalent antibody derivative (F$_{ab}$) produced by reduction (SH) and alkylation of the (F'$_{ab}$)$_2$ species did not generate a cellular response. Nonetheless, crosslinking of the receptor-bound F$_{ab}$ fragments by bivalent anti-F$_{ab}$ molecules yield a cellular response. Results akin to these have been obtained via a totally different approach using derivatives of EGF-URO and of LHRH (see text). A role for receptor microclustering in generating a cell response has thus been postulated[24].

to the EGF-URO receptor but was not mitogenic, became mitogenic when aggregated by anti-EGF-URO antibody[49]. In other work, studying the action of luteinizing hormone releasing hormone (LHRH, also called gonadotropin-releasing hormone or GnRH), two independent laboratories[8, 16, 20] have observed that LHRH antagonists (biologically nonstimulatory derivatives that bind to the receptor and that block the LH-releasing activity of intact LHRH) can be caused to stimulate LH release from pituitary cells in the presence of crosslinking antibodies. The ability of ferritin derivatives of LHRH to cause the microclustering and internalization of LHRH receptors has also been observed at the electron microscopic level[20]. Thus, the microclustering event, associated with cell activation, is probably involved in the activation of a variety of cells by a number of hormones, including insulin.

In the adipocyte, the exact relationship between insulin-mediated receptor microclustering and cell activation is not entirely clear. Since the initial work with antiinsulin receptor antibodies[21, 24] monoclonal antibodies directed against the human insulin receptor have been developed[42]. Because of their nature, the monoclonal bivalent IgG antibodies are thought to react with a single receptor locus, whereas the polyclonal IgG antibody preparations previously used (derived both from insulin-resistant patients and from receptor-injected rabbits) would presumably contain IgG molecules capable of interacting at several receptor loci. In contrast to the polyclonal antibodies for which the results were described above, one of the monoclonal anti-insulin receptor antibody preparations that has been described[42] blocks insulin receptor binding, but does not itself possess intrinsic insulin-like activity in human adipocytes. The lack of response to the monoclonal antibody, which can crosslink the insulin receptors into dimers, is not well understood. Possibly, receptor aggregates containing more than two insulin receptors are required for signalling. Further work will be required to resolve the apparent discrepancies between the observations with the polyclonal and monoclonal antibodies, in terms of the requirement of insulin receptor microclustering for adipocyte activation. Other observations that relate to the question of receptor microclustering and adipocyte activation have been summarized[50]. In brief, studies with insulin-ferritin reveal that a substantial proportion of adipocyte insulin receptors appear to be present in groups of two or more prior to insulin binding. Furthermore, the presence of insulin-ferritin does not lead to an increase in the degree of receptor clustering. These results can be contrasted with the effect of ferritin-LHRH, which clearly caused the microaggregation of previously dispersed LHRH receptors in anterior pituitary cells[20]. Thus, although receptor microaggregation may be a general phenomenon associated with cell activation, the exact mechanism whereby the preclustered insulin receptors in adipocytes are triggered upon occupation by either polyclonal anti-receptor antibody or by insulin (or the insulin-ferritin derivative) remains to be determined. Taken together, the data suggest that the receptor may not only have to be clustered but that in addition, the receptor may also require a special conformational perturbation (e.g., caused only be insulin or by a subset of antireceptor antibodies) to generate the reaction that leads to cell activation. In summary, receptor clustering may be necessary but not necessarily sufficient to initiate a response in target cells for insulin and other hormones.

Internalized receptor and cell activation

It is recognized that agents like insulin or EGF-URO can cause a wide spectrum of cellular responses, ranging in time course from the immediate (seconds to minutes) stimulation of membrane transport (e.g. glucose, amino acids) to the much delayed (hours to tens of hours) effects on DNA synthesis and cell division. How, one may ask, might the receptor dynamics discussed in this article bear on the varied time courses of the multiple actions of insulin, EGF-URO and other hormones? In part, the answer to this question comes from work in fibroblast cell culture systems with the mitogenic/acid-inhibitory polypeptide, epidermal growth factor-urogastrone (EGF-URO)[5, 6, 18, 45, 46]. Under normal circumstances, EGF-URO can, like insulin, activate a large number of cell responses that occur over periods of seconds to minutes (e.g., stimulation of membrane transport and inhibition of acid secretion), up to tens of hours (e.g., stimulation of RNA and DNA synthesis and cell division). A chemical derivative of mouse EGF-URO has been prepared in which the molecule has been cleaved at methionine residue 21 using cyanogen bromide (CNBr). This cleavage results in an opening of one of the disulfide-maintained loops of the EGF-URO molecule. The derivative, denoted CNBr-EGF, still binds to the receptor, but is unable to stimulate DNA synthesis and cell division[49]. Some of the results obtained using CNBr-EGF have already been discussed above. It has also been observed that although microclustering presumably still occurs when CNBr-EGF binds to the receptor, the CNBr-EGF derivative is unable to cause gross aggregation of the EGF-URO receptor. Nonetheless, the nonmitogenic CNBr-EGF derivative is still able to simulate a number of the rapid cellular responses caused by intact EGF-URO (e.g., stimulation of ion flux and induction of morphological changes). The data obtained with CNBr-EGF have been considerably amplified by work with monoclonal antibodies directed against the EGF-URO receptor[45, 46]. In brief, those antibodies capable of causing receptor aggregation (and, consequently, internalization) were able to mimic both the short-term and well as the long-term (mitogenesis) effects of EGF-URO. However, monovalent Fab antibody derivatives that were not capable of inducing receptor aggregation and that were unable to stimulate DNA synthesis, were nonetheless capable of stimulating an early event (membrane phosphorylation) associated with EGF-URO action. A number of hypotheses have thus been put forward concerning the possible role(s) of aggregated receptor and of internalized receptor or internalized receptor fragments in the long-term processes of cell growth or cell differentiation[27]. The short-term membrane-localized reactions may be related to the receptor microclustering process.

As discussed above, it is now realized that in many cell types appreciable amounts of both a ligand-like insulin and its receptor can be taken up and degraded by insulin-responsive cells. Furthermore, there is evidence that binding sites for insulin can be found on Golgi elements, on smooth and rough endoplasmic reticulum[37], as well as on the nuclei of selected cells. Thus, it has been suggested that an internalized degradation product either of insulin or of the insulin receptor may act at an intracellular site to cause at least some of the biological responses triggered by insulin. A role for internalized insulin or insulin fragments appears to be ruled out, in view of the remarkable effects of the anti-receptor antibodies and in view of the stimulatory actions of macro-molecular insulin derivatives (e.g., insulin-Sepharose) that cannot enter the cell. Nonetheless, it is quite possible that there is a role for internalized receptor in directing some of the delayed actions of insulin (e.g., gene regulation and stimulation of cell division). Similarly, a role for internalized receptors (or receptor fragments) must be considered for agents like EGF-URO.

In keeping with the data that have been described in connection with EGF-URO action, it is possible to speculate on a series of events that may occur during the course of insulin-mediated cell activation. Although very hypothetical, the scheme proposed here (fig. 5), derived largely from experiments done with cultured fibroblasts, places in a feasible context both the receptor dynamics described in this article and the intrinsic kinase activities of the insulin and EGF-URO receptors that have recently been described in work from a number of laboratories. (The insulin and EGF-URO receptors can become phosphory-lated and may also phosphorylate other cellular substrates.) It must also be kept in mind that the scheme presented is a generalized picture that synthesizes information obtained from a number of cultured cell systems using several peptide hormone probes. For a specific ligand such as insulin acting on a particular cell type (e.g., the adipocyte), some but not all of the processes outlined in the generalized scheme may in fact occur.

It is quite likely that the rapid actions (seconds to minutes) of insulin, EGF-URO and other agents are caused by biochemical reactions in which the receptor participates within the plane of the plasma membrane. The initial receptor microclustering event probably represents a critical process that is directly related to these membrane-localized reactions. One exciting possibility is that the intrinsic tyrosine kinase activity of the receptors may initiate some of these key reactions. As indicated above, the rapid modulation of cellular activity may require a special receptor conformation that would permit effective interactions between the receptors that associate in microclusters. In this context, the ligands (insulin, EGF-URO) can be viewed as essential allosteric regulators that can modulate membrane-localized reactions (and thereby cell response) on a minute-to-minute basis. Cells would thus be very responsive to extracellular changes in hormone concentrations.

Once internalized, however, the receptor would no longer be exposed to the variations in ligand concentrations that might occur in the extracellular milieu. Thus, the time course of reactions in which the internalized receptor might participate could differ considerably from the time frame of those reactions occurring in the plasma membrane. One role for the internalized receptor may be to regulate some of the delayed effects (hours to tens of hours) that a ligand may have on its target cells.

As hypothesized in figure 5, the internalized receptor-bearing endosome or 'receptosome'[36] may do more than simply function as a way station for the receptor, en route either to the lysosome or, via recycling, back to the cell surface. Rather, the endosome may function as a site-directed receptor-kinase-bearing vesicle that may regulate enzyme activity by phosphorylation reactions that could occur in regions quite distant from the cell surface. In addition, by membrane fusion, the endosome may even transfer the receptor to a new membrane environment, such as the nuclear membrane. Since the endosome-

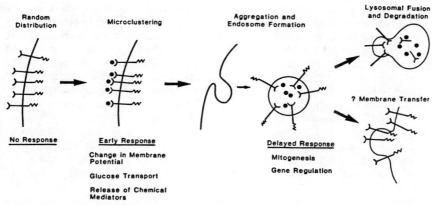

Figure 5. Receptor dynamics and cell activation. As discussed in the text, cell respones that are rapidly regulated probably involve the initial microclustering event. Delayed effects may be caused by a receptor that is internalized in the endosomal organelle. The topography of the endosome would permit the intracellular portion of the receptor (zigzag line) to interact with a variety of intracellular constituents located at considerable distances from the plasma membrane. In the course of its intracellular migration, the receptor-bearing endosome could ultimately fuse either with the lysosomes or with other membrane structures (e.g., Golgi elements or nuclear membranes), resulting in a further relocation of the receptor.

associated receptors are subject to inactivation by lysosomal degradation, a constant influx of fresh receptor-bearing endosomes sustained over a prolonged time period may be required to bring about some of the delayed cellular effects of a ligand like EGF-URO or insulin. The essence of the above discussion is that the temporally distinct actions of certain ligands may relate directly to the topographically distinct receptor dynamic events that, subsequent to ligand binding, occur over quite different time frames. In this context, the continued internalization of receptor may play a key role in the generation of the delayed effects of insulin, EGF-URO and other active ligands. Thus, transmembrane signalling would have two 'tiers' and two associated time frames.

Summary

Although this paper has dealt with general mechanisms whereby a hormonal signal is transmitted across the cell membrane, advances in work with a number of receptors should permit a precision of description of these mechanisms that would have delighted both Langley and Ehrlich. For instance, the detailed sequences now known for the separate subunits of the nicotinic cholinergic receptor[9, 30, 34, 52] and the cellular manipulations made possible by the cloning of the separate subunit genes[31, 51] will make it possible to determine the precise receptor sequence involved either in acetylcholine binding or in ion channel function.The complete sequences and biochemical properties now known for the insulin and EGF-URO receptors[13, 53, 54] to be dealt with in part by a subsequent article (van Obberghen and Gammeltoft, this series) should lay the groundwork for elucidating the transmembrane signalling mechanisms used by the kinase family of growth factor receptors. Continuing work on the structure of the β-adrenergic receptor[29], and on the interaction of such receptors with guanine nucleotide regulatory complexes and on the detailed properties of the family of so-called G-proteins and their associated regulatory subunits[15, 35, 39] should unravel the details for a variety of transmembrane signalling reactions. Thus, at least for three basic transmembrane signalling mechanisms: ligand modulated ion transport; ligand-modulated receptor enzyme activity (e.g. tyrosine kinase); and ligand-modulated liberation of cryptic mediators (like the α- and β-subunits of the guanine nucleotide regulatory complexes) one can look forward with excitement to the elucidation in the not-too-distant future of a number of specific biochemical reaction pathways that lead to cell activation.

1   Berrige, M.J., Inositol triphosphate and diacylglycerol as second messengers. Biochem. J. *220* (1984) 345–360.
2   Boeynaems, J.M., and Dumont, J.E., The two-step model of ligand-receptor interaction. Molec. cell. Endocr. *7* (1977) 33–47.
3   Brown, M.S., and Goldstein, J.L., Lipoprotein receptors in the liver. Control signals for plasma cholesterol traffic. J. clin. Invest. *72* (1983) 743–747.
4   Brown, M.S., Kovanen, P.T., and Goldstein, J.L., Regulation of plasma cholesterol by lipoprotein receptors. Science *212* (1981) 628–635.
5   Carpenter, G., Epidermal growth factor, in: Tissue Growth Factors, Handbook of Experimental Pharmacology, pp. 89–132. Ed. R. Baserga. Springer Verlag, New York 1981.
6   Carpenter, G., and Cohen, S., Epidermal growth factor. A. Rev. Biochem. *48* (1979) 193–216.
7   Catterall, W.A., The emerging molecular view of the sodium channel. Trends Neurosci. *5* (1982) 303–306.
8   Conn, P.M., Rogers, D.C., Stewart, J.M., Neidel, J., and Sheffield, T., Conversion of a gonadotropin-releasing hormone antagonist to an agonist. Nature, Lond. *296* (1982) 653–655.
9   Conti-Tronconi, B.M., and Raftery, M.A., The nicotinic cholinergic receptor: Correlation of molecular structure with functional properties. A. Rev. Biochem. *51* (1982) 491–530.
10   Cuatrecasas, P., Membrane receptors. A. Rev. Biochem. *43* (1974) 169–214.
11   Cuatrecasas, P., and Hollenberg, M.D., Membrane receptors and hormone action. Adv. Prot. Chem. *30* (1976) 251–451.
12   DeHaen, C., The non-stoichiometric floating receptor model for hormone-sensitive adenylate cyclase. J. theor. Biol. *58* (1976) 383–400.
13   Ebina, Y., Ellis, L., Jarnagin, K., Edery, M., Graf, L., Clauser, E., Ou, J.-H., Maslarz, F., Kan, Y.W., Goldfine, I.D., Roth, R.A., and Rutter, W.J., The human insulin receptor cDNA: The structural basis for hormone-activated transmembrane signalling. Cell *40* (1985) 747–758.
14   Frye, L.D., and Edidin, M., The rapid intermixing of cell surface antigens after formation of mouse-human heterokaryons. J. Cell Sci. *7* (1970) 319–335.
15   Gilman, A.G., Guanine nucleotide-binding regulatory proteins and dual control of adenylate cyclase. J. clin. Invest. *73* (1984) 1–4.
16   Gregory, H., Taylor, C.L., and Hopkins, C.R., Leuteinizing hormone release from dissociated pituitary cells by dimerization of occupied LHRH receptors. Nature *300* (1982) 269–271.
17   Hollenberg, M.D., Membrane receptors and hormone action. Pharmac. Rev. *30* (1979) 393–410.
18   Hollenberg, M.D., Epidermal growth factor-urogastrone, a polypeptide acquiring hormonal status. Vit. Horm. *37* (1979) 69–110.
19   Hollenberg, M.D., Receptor-mediated phosphorylation reactions. Trends Pharmac. Sci. *3* (1982) 271–273.
20   Hopkins, C.R., Semoff, S., and Gregory, H., Regulation of gonadotropin secretion to the anterior pituitary. Phil. Trans R. Soc. *B296* (1981) 73–81.
21   Jacobs, S., Chang, K.-J., and Cuatrecasas, P., Antibodies to purified insulin receptor have insulin-like activity. Science *200* (1978) 1283–1284.

22 Jacobs, S., and Cuatrecasas, P., The mobile receptor hypothesis and polypeptide hormone action, in: Polypeptide Hormone Receptors, pp. 39–60. Ed. B. Posner. Marcel Dekker, Inc., New York 1985.

23 Jarett, L., and Seals, J. R., Pyruvate dehydrogenase activation in adipocyte mitochondria by an insulin-generated mediator from muscle. Science 206 (1979) 1407–1408.

24 Kahn, C. R., Baird, K. L., Flier, J. S., Grunfeld, C., Harmon, J. T., Harrison, L. C., Karlsson, F. A., Kasuga, M., King, G. L., Lang, U. C., Podskalny, J. M., and Van Obberghen, E., Insulin receptor, receptor antibodies and the mechanism of insulin action. Recent Prog. Horm. Res. 37 (1981) 477–538.

25 Kaplan, J., Polypeptide-binding membrane receptor: Analysis and classification. Science 212 (1981) 14–20.

26 Kasuga, M., Zick, Y., Blithe, D. L., Grettaz, M., and Kahn, C. R., Insulin stimulates tyrosine phosphorylation of the insulin receptor in a cell-free system. Nature, Lond. 298 (1982) 667–669.

27 King, A. C., and Cuatrecasas, P., Peptide hormone-induced receptor mobility, aggregation and internalization. New Engl. J. Med. 305 (1981) 77–88.

28 Larner, J., Galasko, G., Cheng, K., DePaoli-Roach, A. A., Huang, L., Daggy, P., and Kellogg, J., Generation by insulin of a chemical mediator that controls phosphorylation-dephosphorylation. Science 206 (1979) 1408–1410.

29 Lefkowitz, R. J., Stadel, J. M., and Caron, M. G., Adenylate cyclase-coupled beta-adrenergic receptors: Structure and mechanisms of activation and desensitization. A. Rev. Biochem. 52 (1983) 159–186.

30 Marx, J. L., Cloning the acetylcholine receptor genes. Science 219 (1983) 1055–1056.

31 Mishina, M., Tobimatsu, T., Imoto, K., Tanaka, K.-I., Fujita, Y., Fukuda, K., Kurasaki, M., Takahashi, H., Morimoto, Y., Hirose, T., Inayama, S., Takahashi, T., Kuno, M., and Numa, S., Location of functional regions of acetylcholine receptor -subunit by site-directed mutagenesis. Nature 313 (1985) 364–369.

32 Nexo, E., and Hollenberg, M. D., Characterization of the particulate and soluble acceptor for transcobalamin II from human placenta and rabbit liver. Biochim. biophys. Acta 628 (1980) 190–200.

33 Nishizuka, Y., The role of protein kinase C in cell surface signal transduction and tumour promotion. Nature 308 (1984) 693–698.

34 Noda, M., Takahashi, H., Tanabe, T., Toyosato, M., Kikyotani, S., Furutani, Y., Hirose, T., Takashima, H., Inayama, S., Miyata, T., and Numa, S., Structural homology of Torpedo californica acetylcholine receptor subunits. Nature 302 (1983) 528–532.

34a Noda, M., Shimizu, S., Tanabe, T., Takai, T., Kayano, T., Ikeda, T., Takahashi, H., Nakayama, H., Kanaoka, Y., Minamino, N., Kanagawa, K., Matsuo, H., Raftery, M. A., Hirose, T., Inayama, S., Hayashida, H., Miyata, T., and Numa, S., Primary structure of Electrophorus electricus sodium channel deduced from cDNA sequence. Nature 312 (1984) 121–127.

35 Northup, J. K., Overview of the guanine nucleotide regulatory systems, N_s and N_i, which regulate adenylate cyclase activity in plasma membranes, in: Molecular Mechanisms of Transmembrane Signalling, pp. 91–116. Eds P. Cohen and M. Houslay. Elsevier Science Publishers B.V. (Biomedical Division), Amsterdam 1985.

36 Pastan, I. H., and Willingham, M. C., Journey to the center of the cell: Role of the receptosome. Science 214 (1981) 504–509.

37 Posner, B. I., Khan, M. N., and Bergeron, J. J. M., Receptor-mediated uptake of peptide hormones and other ligands, in: Polypeptide hormone receptors, pp. 61–90. Ed. B. I. Posner. Marcel Dekker, Inc., New York 1985.

38 Raff, M., Self regulation of membrane receptors. Nature 259 (1976) 265–266.

39 Rodbell, M., The role of hormone receptors and GTP-regulatory proteins in membrane transduction. Nature 284 (1980) 17–22.

40 Rosen, O. M., Herrera, R., Olowe, Y., Petruzzelli, L. M., and Cobb, M. H., Phosphorylation activates the insulin receptor tyrosine protein kinase. Proc. natn. Acad. Sci. USA 80 (1983) 3237–3240.

41 Roth, R. A., and Cassell, D. J., Evidence that the insulin receptor is a protein kinase. Science 219 (1983) 299–301.

42 Roth, R. A., Cassell, D. J., Wong, K. Y., Maddux, B. A., and Goldfine, I. D., Monoclonal antibodies to the human insulin receptor block insulin binding and inhibit insulin action. Proc. natn. Acad. Sci. USA 79 (1982) 7312–7316.

43 Saltiel, A. R., Siegel, M. I., Jacobs, S., and Cuatrecasas, P., Putative mediators of insulin action: Regulation of pyruvate dehydrogenase and adenylate cyclase activities. Proc. natn. Acad. Sci. USA *79* (1982) 3513–3517.

44 Schlessinger, J., Shechter, Y., Cuatrecasas, P., Willingham, M. C., and Pastan, I., Quantitative determination of the lateral diffusion coefficients of the hormone-receptor complexes of insulin and epidermal growth factor on the plasma membrane of cultured fibroblasts. Proc. natn. Acad. Sci. USA *75* (1978) 5353–5357.

45 Schlessinger, J., Schreiber, A. B., Levi, A., Lax, I., Libermann, T. A., and Yarden, Y., Regulations of cell proliferation by epidermal growth factor. CRC Crit. Rev. Biochem. *14* (1983) 93–112.

46 Schreiber, A. B., Libermann, T. A., Lax, I., Yarden, Y., and Schlessinger, J., Biological role of epidermal growth factor-receptor clustering. J. biol. Chem. *258* (1983) 846–853.

47 Seals, J. R., and Czech, M. P., Evidence that insulin activates an intrinsic plasma membrane protease in generating a secondary chemical mediator. J. biol. Chem. *255* (1980) 6529–6531.

48 Seligman, P. A., and Allan, R. H., Characterization of the receptor for transcobalamin II isolated from human placenta. J. biol. Chem. *253* (1978) 1766–1772.

49 Shechter, Y., Hernaez, L., Schlessinger, J., and Cuatrecasas, P., Local aggregation of hormone-receptor complexes is required for activation by epidermal growth factor. Nature, Lond. *278* (1979) 835–838.

50 Smith, R. L., and Jarett, L., Tissue specific variations in insulin receptor dynamics: a high resolution ultrastructural and biochemical approach, in: Insulin: Its Receptor and Diabetes, pp. 105–139. Ed. M. D. Hollenberg. Marcel Dekker, Inc., New York 1985.

50a Sokolovsky, M., Muscarinic receptors in the central nervous system. Int. Rev. Neurobiol. *25* (1984) 139–183.

51 Stevens, C. F., Molecular tinkerings that tailor the acetylcholine receptor. Nature *313* (1985) 350–351.

52 Taylor, P., Brown, R. D., and Johnson, D. A., The linkage between ligand occupation and response of the nicotinic acetylcholine receptor, in: Current topics in Membranes and Transport, vol. 18, pp. 407–444. Ed. A. Kleinzeller. Academic Press, New York 1983.

53 Ullrich, A., Coussens, L., Hayflick, J. S., Dull, T. J., Gray, A., Tam. A. W., Lee, J., Yarden, Y., Libermann, T. A., Schlessinger, J., Downward, J., Mayes, E. L. V., Whittle, N., Waterfield, M. D., and Seeburg, P. H., Human epidermal growth factor receptor cDNA sequence and aberrant expression of the amplified gene in A431 epidermoid carcinoma cells. Nature *309* (1984) 418–425.

54 Ullrich, A., Bell, J. R., Chen, E. Y., Herrera, R., Petruzzelli, L. M., Dull, T. J., Gray, A., Coussens, L., Liao, Y.-C., Tsubokawa, M., Mason, A. Seeburg, P. H., Grunfeld, C., Rosen, O. M., and Ramachandran, J., Human insulin receptor and its relationship to the tyrosine kinase family of oncogenes. Nature *313* (1985) 756–761.

55 Van Obberghen, E., and Kowalski, A., Phosphorylation of the hepatic insulin receptor. Stimulating effect of insulin on intact cells and in a cell-free system. FEBS Lett. *143* (1982) 179–182.

56 Zierler, K., Membrane polarization and insulin action, in: Insulin: Its Receptor and Diabetes, pp. 141–179. Ed. M. D. Hollenberg. Marcel Dekker, Inc., New York 1985.

57 Zierler, K., and Rogus, E. M., Effects of peptide hormones and adrenergic agents on membrane potentials of target cells. Fedn Proc. *40* (1981) 121–124.

Insulin receptors: Structure and function

E. Van Obberghen and S. Gammeltoft

INSERM U 145, Faculté de Médecine (Pasteur), Université de Nice, Avenue de Vallombrose, F–06034 Nice Cedex (France), and Dept Clinical Chemistry, Rigshospitalet, University Hospital, Blegdamsvej 9, DK–2100 Copenhagen (Denmark)

Introduction

More than fifty years after the discovery of insulin, its cellular mechanism of action still remains one of the major obstacles in cell biology. Recent progress in the molecular characterization of the insulin receptor itself due to concerted efforts in several laboratories have led to important discoveries. One of these, the kinase activity and autophosphorylation of the insulin receptor, will be reviewed and its putative role in insulin action discussed. Furthermore, conditions with cellular insulin resistance are coupled with decreased phosphorylation of the insulin receptor, giving a clue to a molecular defect in the disease states.

The molecular mechanism of insulin action

Regulation of cellular metabolism and growth by insulin is a result of a series of events initiated by the interaction of the hormone with specific cell surface receptors (fig. 1). In the past, insulin receptors on a large number of cell types have been characterized in detail by their structure and function[33, 38, 75]. This achievement is based on the development and application of a variety of biochemical methods including kinetic analysis for description of the receptor binding[16]; affinity labeling technique for identification of the receptor subunits[10]; and recently, recombinant DNA technology for the elucidation of receptor amino acid sequences[13, 73]. In spite of this progress, the molecular mechanism of insulin action is still poorly comprehended as far as the events following the receptor binding and leading to the ultimate cellular responses are concerned. Many attempts to isolate a second messenger in insulin action have proved this to be difficult[10, 16, 33, 38, 75].

Recently, a promising discovery was made when it was demonstrated that the insulin receptor is an insulin-sensitive protein kinase[1, 24, 39, 42, 50, 60, 66, 78, 79]. This novel observation is of interest for our understanding of insulin-regulated processes, since it is now recognized that covalent phosphorylation-dephosphorylation of proteins is a mechanism whereby many cellular functions are regulated by hormones and neurotransmitters[7, 11]. Furthermore, protein kinases are also constituents of receptors for several polypeptide growth factors includ-

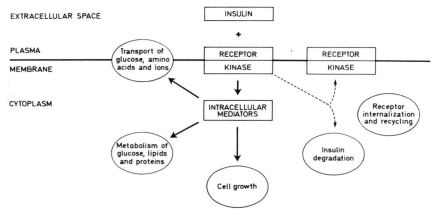

EXTRACELLULAR SPACE

PLASMA

MEMBRANE

CYTOPLASM

Figure 1. Cellular mechanism of insulin action. The receptor-kinase complex in the plasma membrane transmits the intracellular insulin signal to intracellular mediators e.g. phosphoproteins, which stimulate transport of glucose, amino acids and ions, metabolism of glucose, lipids and proteins and cell growth. The receptor-bound insulin is internalized and degraded whereas the receptor is recycled to the plasma membrane.

ing epidermal growth factor (EGF)[8], platelet-derived growth factor (PDGF)[14], transforming growth factor (TGF-α)[57], and insulin-like growth factor-I (IGF-I)[35, 64], implying that receptor kinase activity may represent a general mechanism in transmembrane signalling of hormones and growth factors.

The insulin receptor kinase

The insulin receptor is an integral membrane glycoprotein ($M_r \sim 350,000$) composed of two α-subunits ($M_r \sim 130,000$) and two β-subunits ($M_r \sim 95,000$) linked by disulfide bonds[33, 38, 75] (fig. 2). Affinity labeling of the receptor using

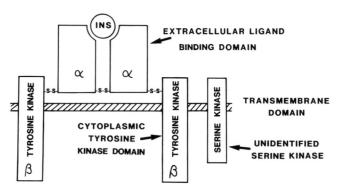

Figure 2. Schematic model of the insulin receptor kinase complex.

either photosensitive insulin analogues[34, 82], or cross-linking of insulin with bifunctional reagents[53, 68] have shown that the α-subunit is labeled predominantly by radioactive insulin, when compared to the β-subunit, the labeling of which is much weaker[47, 77], or even absent[82]. This suggests that the insulin binding site is located on the α-subunit of the receptor oligomer.

In intact cells, insulin stimulates the phosphorylation of its receptor β-subunit. This was first demonstrated in rat hepatoma cells and human IM-9 lymphoblasts[39], and later in freshly isolated rat hepatocytes[78]. In these experiments, cells were preincubated with ^{32}P-ortho-phosphate to label cellular ATP, solubilized in detergent, and the glycoproteins purified on wheat-germ-agglutinin-agarose. Immunoprecipitation of phosphorylated proteins by antibodies to insulin receptor followed by sodium-dodecyl-sulfate (SDS) polyacrylamide gel electrophoresis under reducing conditions and autoradiography revealed a labeled band ($M_r \sim 95,000$), the phosphorylation of which was stimulated by insulin. Its identity with the insulin receptor β-subunit was established for the following reasons. First, non-immune serum did not precipitate a band with a similar electrophoretic mobility. Second, the molecular size was identical with that determined previously, using biosynthetic and affinity labeling methods[28, 47, 53, 68, 77].

Subsequently, the phosphorylation of the β-subunit of the insulin receptor was demonstrated in cell-free systems using [γ-^{32}P] ATP in solubilized and partially purified receptor preparations from rat liver (fig. 3) and human placenta[1, 24, 42, 50, 60, 66, 78, 79]. Phosphoaminoacid analysis of the phosphorylated β-subunit of partially purified receptors showed phosphoserine, phosphothreonine and phosphotyrosine under basal conditions. Insulin induced a several-fold increase in ^{32}P-incorporation in tyrosine, and had in addition a smaller, but consistent stimulating effect on the labeling of phosphoserine[19, 43].

The insulin receptor exhibits insulin-dependent tyrosine-kinase activity. This was demonstrated in cell-free systems with detergent-solubilized and highly purified receptors obtained from various tissues. The purification scheme was based on sequential affinity chromatography on wheat-germ-agglutinin- and insulin-agarose[15, 34, 40, 41, 48, 51]. Alternatively, the lectin-purified receptors were immunoprecipitated with antibodies to insulin receptor, obtained from patients with severe insulin-resistance and Acanthosis nigricans[27, 42, 43, 79], or monoclonal IgG directed against insulin receptors[45, 59, 63]. These purified receptor preparations exhibited insulin-stimulated protein kinase activity which catalyzed phosphorylation of both the β-subunit and exogenous substrates like casein, histones and synthetic tyrosine-containing peptides[1, 19, 41, 43, 50, 51, 60, 66, 79]. In contrast to the partially purified receptor, the phosphorylation occurred exclusively on tyrosine residues in highly purified receptors under basal conditions, and the insulin stimulatory action was accounted for by a several-fold increase in phosphotyrosine[1, 19, 43, 50, 60, 66]. Thus, the tyrosine kinase is a constituent of the insulin receptor, whereas the serine kinase is non-covalently associated with the receptor (fig. 2). In addition to being the principal substrate for autophosphorylation, the β-subunit has an ATP-binding site as demonstrated by covalent affinity labeling with oxidized [α-^{32}P]ATP[79] or photoreactive azido [α-^{32}P]ATP[66]. The simultaneous presence of a phosphorylation site and an ATP-binding site on the β-subunit suggests that the insulin receptor acts as its own

34

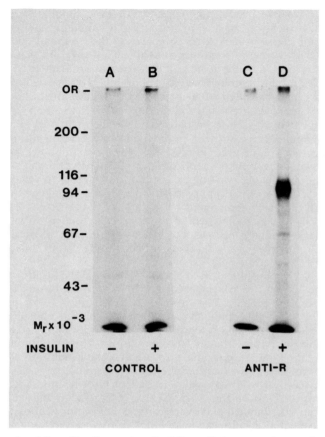

Figure 3. Phosphorylation of insulin receptors. Partially purified receptors from rat liver were incubated 30 min at 20°C in the absence (A and C) or presence of insulin (10^{-7} mol/l) (B and D) and phosphorylated with [γ-^{32}P]ATP. The phosphoproteins were immunoprecipitated with anti-insulin receptor antiserum (C and D) or normal serum (A and B) and analyzed by sodium dodecyl sulfate polyacrylamide gel electrophoresis followed by autoradiography[78].

tyrosine kinase. Further proof of the identity of the insulin receptor kinase seems to be the demonstration that the insulin-binding activity and the insulin-dependent tyrosine-kinase activity co-purified at a constant stoichiometric ratio to homogeneity[40, 41, 48, 51, 60]. Thus, the functional cell-surface insulin receptor is composed of two functional domains, one with binding activity and another with tyrosine-specific protein kinase activity. In addition, insulin binds to and promotes phosphorylation of the insulin receptor precursor, a monomeric protein of $M_r \sim 210,000$[3, 54].

Biochemical properties of the insulin receptor kinase

Following the identification of the protein kinase activity of the insulin receptor β-subunit, its biochemical properties were investigated (table 1). These included temperature dependence, metal ion requirements, nucleotide and substrate specificity, and kinetic parameters of the phosphorylation reaction. In the absence of insulin phosphorylation occurred slowly, but addition of insulin (100 nM) rapidly stimulated the incorporation of ^{32}P from $[\gamma-^{32}$P]ATP into the β-subunit of the receptor. Within 30 s at 22 °C, autophosphorylation of the insulin-stimulated receptor reached 50 % of maximum and a steady state value was reached after about 10 min[81]. Even at 4 °C, the phosphorylation was rapid; the ^{32}P-content of the receptor reached half-maximal level by 5 min and maximum after about 20 min[88].

As with the EGF-stimulated phosphorylation of the EGF-receptor[6, 52], Mn^{2+} was the most potent cation in augmenting the insulin-stimulated phosphorylation of the insulin receptor[1,48,51,52,81,88]. The effect of Mn^{2+} was maximal at concentrations above 2 mM and constant up to 10 mM[52], but showed a complex relationship with the ATP-concentration (see below). Mg^{2+} was also effective, but concentrations above 15 mM were required for a maximal effect. However, the insulin-stimulated kinase showed greater activity in the presence of a combination of 2 mM Mn^{2+} and 12 mM Mg^{2+} than when either metal ion was used alone[43]. Ca^{2+} as well as Zn^{2+} and Cr^{2+} were totally ineffective, whereas Co^{2+} (2 mM) had some effect[1, 88]. This ion dependency is characteristic of tyrosine kinases compared to serine kinases and threonine kinases[8].

In the cell-free system, the source of phosphate used to phosphorylate the β-subunit was identified as ATP[42], and the K_m value for ATP of the insulin-stimulated receptor kinase was determined as 30–150 μM[48, 51, 52, 81]. No other nucleotide triphosphate (GTP, CTP, TTP or UTP) competed with $[\gamma-^{32}$P]ATP in the

Table 1. Major features of the insulin receptor kinase*

1. Intrinsic to the receptor
   - ATP-binding site on the receptor β-subunit
   - Phosphorylation of highly purified receptors
   - Co-purification of insulin binding activity and insulin-stimulated kinase activity
   - Present when receptors present

2. Regulators
   - Insulin
   - ATP (phosphate donor)
   - Mn^{2+}, Mg^{2+}

3. Substrates
   - Receptors: autophosphorylation
   - Substrates: exogenous and endogenous

4. Phosphoamino acids in receptor
   - Intact cells: tyrosine and serine
   - Cell-free systems: predominantly tyrosine

5. Multiple sites phosphorylated on receptor β-subunit

*The features listed were compiled from data in refs. 1, 41–43, 48, 50–52, 58, 60, 66, 70, 79, 81, 83, 84, 88.

receptor phosphorylation assay, whereas addition of ATP and ADP, but not AMP, gave significant inhibition of ^{32}P incorporation[48, 52, 81]. Cyclic AMP had no effect on the phosphorylation of the receptor[39, 48, 50]. Thus, the insulin receptor kinase showed specificity for adenosine di- and tri-nucleotides. As mentioned above, Mn^{2+} and ATP showed a complex relationship in their activation of the kinase. Kinetic data showed that Mn^{2+} acted predominantly by decreasing the K_m for ATP presumably through binding to a specific regulatory site on the kinase rather than chelating with ATP. On the other hand, increasing ATP concentration decreased the K_m for Mn^{2+}, showing that a high substrate concentration can activate the kinase even when the metal activator concentration is low[81].

The substrate specificity of the insulin receptor kinase was assessed using both naturally occurring proteins and synthetic peptides including histones, casein, tubulin, troponin, angiotensin II, angiotensin II inhibitor, β-lipotropin, pp60src (a gene product of the Rous sarcoma virus), anti-pp60src IgG, and several synthetic peptide fragments containing a tyrosine residue[19, 41, 48, 52, 70]. In the proteins, phosphoaminoacid analysis showed only phosphorylation on tyrosine residues. Among the synthetic peptides, even a dipeptide, Tyr-Arg was a substrate although with very high K_m[70]. The K_m values varied significantly among the substrates from 1 μM to a value > 80 mM, but insulin acted by stimulating the V_{max} with no alteration of K_m[41, 48, 52, 70]. The substrate specificity of the insulin receptor kinase was similar, but not identical with that of the EGF receptor[46, 52] and the pp60src kinases[30], suggesting that they are members of a superfamily of tyrosine kinases which has diverged from a common evolutionary origin.

It seems that the insulin receptor β-subunit is the best substrate for its own kinase. This conclusion is based on the observation that the V_{max} for autophosphorylation of the insulin receptor kinase was increased nearly 20-fold by insulin[81], whereas the V_{max} values for other substrates were only increased 2–5-fold[41, 48, 52, 70]. Alternatively, it has been suggested that the kinase activity associated with the insulin receptor is increased by tyrosine phosphorylation of the receptor β-subunit[58, 83, 84]. Phosphorylation on tyrosine residues induced by insulin leads to increased kinase activity, whereas dephosphorylation of the tyrosine residues by alkaline phosphatase is accompanied by a marked inhibition[83]. Thus, the autophosphorylation on tyrosine residues may play a key role in regulating the insulin receptor kinase. The additional phosphorylation of serine and threonine residues on the receptor by non-covalently associated kinases may also exert a regulatory role on insulin receptor kinase and binding activity[10].

Role of receptor phosphorylation in insulin action

The biological relevance of insulin receptor phosphorylation is not clear. It is possible that it plays a role in cellular processes such as receptor affinity regulation, hormone and receptor internalization and signal transmission. These phenomena are well-characterized[10, 16, 33, 38, 75], but their molecular mechanism is almost completely unknown. Most likely, receptor regulation, internalization and transmembrane signaling are integrated events in insulin action, and receptor autophosphorylation per se is involved in transmission of the insulin message to cellular enzymes and transport carriers. At present, it is tempting to suggest that the covalent receptor modification is an early step in insulin action and that the increased kinase activity of the insulin receptor evoked by hormone binding would lead to phosphorylation-dephosphorylation of other cellular proteins, and through the generation of a cascade of reactions this would result in the final effects of insulin. Five requirements should be fulfilled by the insulin-induced receptor kinase activation and autophosphorylation before one can say with certainty that they are involved in physiological insulin action.

First, the insulin dose-response relationship of the kinase should be within the physiological range and correlate with that of the binding to the receptor. Several authors found that the kinase activation was half-maximal at an insulin concentration of 2–5 nM ($\sim ED_{50}$), which corresponded to the apparent K_d of the receptor-insulin complex as determined with the same preparations of solubilized receptor of varying purity obtained from human placenta[1, 43, 48, 51, 52, 58, 65, 66]. In contrast, a dissociation between the dose-response curves of insulin binding and kinase activation was observed with soluble receptors from rat liver and human erythrocytes[21, 88]. In these studies, the apparent K_d exceeded the ED_{50} by a factor of 3–10, which suggested that the phenomenon of 'spare receptors' observed for other insulin actions[16] is also applicable for kinase activation. It is not clear whether the different findings are the result of differences in the tissues, purification procedures, or assay methods used. In conclusion, the receptor kinase is activated by insulin concentrations within a physiological range corresponding to the receptor binding.

Secondly, the receptor kinase should be capable of phosphorylating cellular substrate other than the receptor itself, in order to propagate the insulin response. As discussed in detail above, the insulin receptor kinase is capable of phosphorylating a number of substrates on tyrosine residues, in vitro, although none of the proteins tested are physiological substrates[19, 41, 51, 52, 70]. Recently, two laboratories, independently, identified a cellular protein substrate of $M_r \sim 110,000$–120,000 for the insulin receptor kinase in wheat-germ-agglutinin purified glycoproteins from rat liver and rabbit brown adipose tissue[56, 65]. The naturally occurring glycoprotein appears as a monomeric structure, and it is not part of the insulin receptor itself, because it was not immunoprecipitated by highly specific antibodies to insulin receptor. Phosphorylation of the $M_r \sim 110,000$ protein and autophosphorylation of the receptor β-subunit ($M_r \sim 95,000$) were stimulated by insulin in a remarkably similar dose-dependent fashion with an ED_{50} of 1 nM. Further kinetic studies suggested that the phosphorylation of the $M_r \sim 110,000$ protein occurred after autophosphorylation of the insulin receptor kinase[65]. The nature and function of this endogenous substrate is as yet unknown; nor can we answer the intriguing question whether it displays kinase or phosphatase activity.

In intact cells, a rapid insulin-stimulated phosphorylation of its receptor on tyrosine residues is followed by a slower serine phosphorylation[43]. Furthermore, in a cell-free system of partially purified receptor, some laboratories have reported that insulin stimulates phosphorylation of both tyrosine and serine residues of its receptor[43, 83, 85], as well as on exogenous substrates[19]. The serine kinase is non-covalently associated with the receptor, and is removed during further purification, because the highly purified receptor displayed only tyrosine kinase activity[19, 41, 42, 48, 51]. The relationship between the two protein kinase activities associated with the receptor and their cellular role remains to be established. Two possibilities exist (fig. 4), one in which both kinases serve separate cellular functions, and another one with sequential activation of the kinases[76].

The first model implies that the tyrosine-specific enzyme activity is involved in insulin's growth-promoting action in a similar way to the tyrosine phosphorylations mediating the cellular responses to growth factors such as EGF[8], PDGF[14] and TGF-α[57] and several cellular and retroviral oncogene proteins[2, 29, 31]. In contrast, the serine kinase would play a role in insulin's metabolic actions. All kinases involved in the control of intermediary metabolism are indeed serine or threonine specific[7, 11], and phosphoserine and phosphothreonine constitute about 99.97% of all phosphorylated amino acids; phosphotyrosine accounts for the remaining 0.03%[32]. In the second model, the two kinases are activated sequentially. Insulin binding to its receptor leads to activation of the constituent tyrosine kinase, which then induces activation of the receptor-associated serine kinase and this last one accounts for the generation of cellular responses to insulin. Future work should be directed towards the identification of the serine kinase.

The third criterion is that of reversibility of insulin receptor phosphorylation. To exert a regulatory function, the phosphorylated and activated receptor kinase should return to basal activity through a dephosphorylation reaction. Solubilized, lectin-purified receptors from rat liver membranes contained phosphatase activity, which slowly reduced the ^{32}P-content of the phosphorylated receptor, and which was insulin-independent[25, 44, 86]. The physiological significance of this reaction is difficult to ascertain, in particular because a 2-h incubation at 22 °C was required for complete dephosphorylation of the recep-

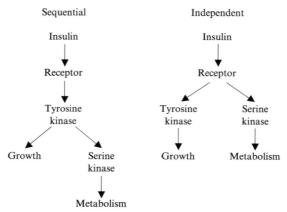

Figure 4. Putative roles of receptor-associated tyrosine and serine kinases in insulin control of cell metabolism and growth.

tor[44]. A different approach was taken by incubation of the phosphorylated insulin receptor with alkaline phosphatase, which resulted in removal of about 50% of the phosphotyrosine in the β-subunit, and about 65% reduction in kinase activity, suggesting that dephosphorylation is accompanied by deactivation of the receptor kinase[83, 86]. These observations demonstrate that the insulin receptor kinase can be deactivated through dephosphorylation of phosphotyrosine residues, although the physiological mechanism remains to be elucidated.

The fourth criterion is the specificity of the insulin effect on its receptor kinase. Several insulin analogues including porcine proinsulin, desoctapeptide insulin, desalanine-desasparagine insulin, guinea pig insulin, insulin-like growth factor II and covalently linked insulin dimers stimulated receptor phosphorylation with potencies relative to porcine insulin which were identical to their relative binding affinities and potencies in other assay systems[21, 42, 50, 52, 62, 88]. Furthermore, polyclonal antisera directed against the insulin receptor, which show insulin-like effects in several cell-types, were also able to stimulate the tyrosine-specific kinase associated with the receptor[50, 85]. However, some antibodies were inactive, although they showed both insulin-like effects in intact cells and interaction with receptors in cell-free preparations[69, 87]. The reason for this discrepancy is not clear, but a possible explanation is that activation of the receptor-associated tyrosine kinase mediates the growth activity of insulin and not the metabolic actions (fig. 4). Finally, other hormones which do not bind to the insulin receptor, including EGF, which activates its own receptor kinase, had no effect on insulin receptor kinase activity[53, 82]. In conclusion, the insulin effect on phosphorylation of its own receptor has the affinity and specificity of a typical insulin receptor mediated event.

Finally, the kinase activity is present whenever insulin receptors are present. So far, receptors in all cell-types investigated have shown insulin-stimulated phosphorylation of the β-subunit. These include liver[42, 78, 79], adipose tissue[24, 80], skeletal muscle[5, 26, 46], placenta[50, 60, 66], lymphocytes[22], erythrocytes[21], fibroblasts[76], brain cortex[17, 55] and various tumor cell lines like IM-9 lymphoblasts[39], 3T3-L1 adipocytes[50], hepatoma[39, 43, 87] and insulinoma cells[18]. Thus, the insulin-sensitive kinase is a general feature of the insulin receptor.

Additional evidence for a role of insulin receptor kinase in insulin action was obtained from two kinds of observations (table 2). First, the receptor kinase

Table 2. Evidence for a role of insulin receptor kinase in insulin action*

Impaired insulin receptor kinase activity in insulin-resistant states
  1. Syndrome of insulin resistance and Acanthosis nigricans type A
  2. Cultured melanoma cells
  3. Mice rendered obese by goldthioglucose
  4. Streptozotocin-diabetic rats

Increased insulin receptor phosphorylation induced by insulinomimetic agents
  1. Vanadate
  2. Trypsin
  3. Concanavalin A
  4. Wheat-germ-agglutinin

* Data from refs. 22, 23, 27, 37, 44, 46, 61, 71, 72.

activity is impaired in cells from various insulin-resistant states including the syndrome of insulin resistance and Acanthosis nigricans, type A[20-22], from melanoma cell cultures[27], from goldthioglucose obese mice[46], and from streptozotocin-diabetic rats[37]. Second, insulinomimetic agents like vanadate, concanavalin A, wheat-germ-agglutinin, and trypsin, which act via the insulin receptor, increased the receptor autophosphorylation[44, 61, 71, 72].

In conclusion, five criteria are fulfilled which establish the kinase activity as a fundamental property of the insulin receptor and strongly suggest an important role in insulin action. Data from several laboratories suggest that receptor phosphorylations are involved in insulin receptor autoregulation and in the transmission of the insulin signal. At present there is no information on a role in insulin receptor internalization. It has been proposed that tyrosine phosphorylation of the β-subunit regulates its kinase activity, whereas receptor phosphorylation on serine and threonine residues could play a role in modulation of the binding affinity of the α-subunit as well as kinase activity of the β-subunit. Furthermore, phosphorylation of an endogenous substrate on tyrosine and serine might represent a secondary event leading to insulin actions on cellular metabolism and growth.

Structure-function relationship of the insulin receptor kinase

Several authors have attempted to purify the insulin receptor for structural analysis. The protocols used were mainly based on affinity chromatography using agarose conjugated with lectins such as concanavalin or wheat-germ-agglutinin, followed by insulin-agarose[9, 15, 36, 41, 48, 51, 63, 67, 68]. At least three laboratories succeeded in purifying the receptor from placental membranes[15, 51, 63]. The pure insulin receptor has a binding capacity of 1.1–1.5 mol insulin per mol of receptor (M_r 300,000)[15, 51], and protein kinase activity with a V_{max} of 80 mmol/min/mg (using angiotensin as substrate)[51].

Recently, the amino acid sequence of the human insulin receptor precursor was deduced from human placental complementary DNA (cDNA) clones[9, 73]. This achievement was based on amino-terminal sequences obtained for both α- and β-subunits of the purified receptor, which were used for the design of single long synthetic DNA probes and hybridization screening of a DNA library to identify human insulin receptor cDNA clones. Nucleotide sequence analysis of cDNA positive clones which hybridized with both α- and β-subunit DNA probes revealed a sequence of 5181 base-pairs which coded for 1382 amino acids, including a 27-residue signal peptide. This is the amino acid sequence of the insulin receptor single chain precursor, composed of a N-terminal α-subunit (735 residues) followed by a β-subunit (620 residues) and an intervening peptide composed of 4 basic amino acids (Arg-Lys-Arg-Arg), which probably represents the cleavage site for the receptor precursor processing enzyme[9, 73]. The α-chain is largely hydrophilic, with a few short hydrophobic stretch and contains sequences for asparagine N-linked glycosylation and an unusually large number of 37 cysteine residues. The β-chain contains a sequence of 23–26 hydrophobic amino acids which probably represents a single transmembrane region dividing the β-subunit into a shorter extracellular portion, which links the α-subunit through disulfide bridges, and a longer cytoplasmic part (fig. 2). The cytoplasmic part of the insulin receptor β-subunit shows some homology with other tyrosine-specific kinases like the *src* oncogene kinases[29, 31] and the EGF-receptor kinase[12, 74]. The similarities in sequence include the ATP-binding site and the residues essential for kinase activity as well as tyrosine residues which can be phosphorylated, demonstrating that the insulin receptor is a member of the *src* family of tyrosine kinases. No cellular proto-oncogene has yet been identified which is identical with the insulin receptor β-subunit as is the case for the EGF receptor and *v-erb B* oncogene product[73, 74], although one region of the insulin receptor β-subunit (51 residues) is practically identical with a portion of the *v-ros* transforming protein[49]. It is possible that the insulin receptor is the cellular homologue of the *v-ros* transforming protein, which has a $M_r \sim 68,000$, tyrosine kinase activity and a hydrophobic transmembrane region at the N-terminus[49].

In conclusion, the amino acid sequence of the insulin receptor gives evidence that the β-subunit is a tyrosine kinase. Future studies will define the phosphorylation site at tyrosine and serine residues which might help in understanding the functional role of receptor phosphorylations.

Summary and conclusions

The recent characterization of the human insulin receptor structure and its intrinsic tyrosine kinase activity represent major advances in our understanding of the mechanism of insulin action. It is reasonable to think that the insulin-induced autophosphorylation and activation of its receptor kinase represent an important event in the action of insulin on cell metabolism and growth. The fundamental research reviewed may be followed by the discovery of molecular receptor defects in clinical syndromes of insulin resistance.

Note added in proof:
Two papers have recently described that in intact cells, insulin induces a rapid several-fold increase in ^{32}P-incorporation in tyrosine of its receptor β-subunit followed by a slower rise in labeling of phosphoserine (Pang, D. T., Sharma, B. R., Schafer, J. A., White, M. F., and Kahn, C. R., Predominance of tyrosine phosphorylation of insulin receptors during the initial response of intact cells to insulin. J. biol. Chem. *260* (1985) 7131–7136; White, M. F., Takayama, S., and Kahn, C. R., Differences in sites of phosphorylation of the insulin receptor in vivo and in vitro. J. biol. Chem. *260* (1985) 9470–9478).

Acknowledgments. We thank G. Visciano for preparing the illustrations. The European Molecular Biology Organization and the Institut National de la Santé et de la Recherche Médicale are thanked for supporting S. G. during a short-term fellowship in INSERM U 145, Nice, France.

1   Avruch, J., Nemenoff, R. A., Blackshear, P. J., Pierce, M. N., and Osathanondh, R., Insulin-stimulated tyrosine phosphorylation of the insulin receptor in detergent extracts of human placental membranes. Comparison to epidermal growth factor-stimulated phosphorylation. J. biol. Chem. *257* (1982) 15162–15166.

2   Bishop, J. M., Cellular oncogenes and retroviruses. A. Rev. Biochem. *52* (1983) 301–354.

3   Blackshear, P. J., Nemenoff, R. A., and Avruch, J., Insulin binds to and promotes the phosphorylation of a M_r 210,000 component of its receptor in detergent extracts of rat liver microsomes. FEBS Lett. *158* (1983) 243–246.

42

4   Blackshear, P. J., Nemenoff, R. A., and Avruch, J., Characteristics of insulin epidermal growth factor stimulation of receptor autophosphorylation in detergent extracts of rat liver and transplantable rat hepatomas. Endocrinology *114* (1984) 141–152.

5   Burant, C. F., Treutelaar, M. K., Landreth, G. E., and Buse, M. G., Phosphorylation of insulin receptors solubilized from rat skeletal muscle. Diabetes *33* (1984) 704–708.

6   Carpenter, G., King, L. Jr, and Cohen, S., Rapid enhancement of protein phosphorylation in A-431 cell membrane preparations by epidermal growth factor. J. biol. Chem. *254* (1979) 4884–4891.

7   Cohen, P., The role of protein phosphorylation in neural and hormonal control of cellular activity. Nature *296* (1982) 613–620.

8   Cohen, S., Carpenter, G., and King, L. Jr, Epidermal growth factor-receptor-protein-kinase interactions, co-purification of receptor and epidermal growth factor-enhanced phosphorylation activity. J. biol. Chem. *255* (1980) 4834–4842.

9   Cuatrecasas, P., Properties of the insulin receptor isolated from liver and fat cells membranes. J. biol. Chem. *247* (1972) 1980–1991.

10   Czech, M. P., New perspectives on the mechanism of insulin action. Recent Prog. Horm. Res. *40* (1984) 347–377.

11   Denton, R. M., Brownsey, R. M., and Belsham, G. J., A partial view of the mechanism of insulin action. Diabetologia *21* (1981) 347–362.

12   Downward, J., Yarden, Y., Scrace, G., Totty, N., Stockwell, P., Ullrich, A., Schlessinger, J., and Waterfield, M. D., Close similarity of epidermal growth factor receptor and *v-erb-B* oncogene protein sequences. Nature *307* (1984) 521–524.

13   Ebina, Y., Ellis, L., Jarnagin, K., Edery, M., Graf, L., Clauser, E., Ou, J.-H., Masiarz, F., Kan, Y. W., Goldfine, I. D., Roth, R. A., and Rutter, W. J., The human insulin receptor cDNA: the structural basis for hormone-activated transmembrane signalling. Cell *40* (1985) 747–758.

14   Ek, B., Westermark, B., Wasteson, A., and Heldin, C.-H., Stimulation of tyrosine-specific phosphorylation by platelet-derived growth factor. Nature *295* (1982) 419–421.

15   Fujita-Yamaguchi, Y., Choi, S., Sakamoto, Y., and Itakura, K., Purification of insulin receptor with full binding activity. J. biol. Chem. *258* (1983) 5045–5049.

16   Gammeltoft, S., Insulin receptors: Binding kinetics and structure-function relationship of insulin. Physiol. Rev. *64* (1984) 1321–1378.

17   Gammeltoft, S., Kowalski, A., Fehlmann, M., and Van Obberghen, E., Insulin receptors in rat brain: insulin stimulates phosphorylation of its receptor beta-subunit. FEBS Lett. *172* (1984) 87–90.

18   Gazzano, H., Halban, P., Prentki, M., Ballotti, R., Brandenburg, D., Fehlmann, M., and Van Obberghen, E., Identification of functional insulin receptor on membranes from an insulin-producing cell line (RINm5F). Biochem. J. *226* (1985) 867–872.

19   Gazzano, H., Kowalski, A., Fehlmann, M., and Van Obberghen, E., Two different protein kinase activities are associated with the insulin receptor. Biochem. J. *216* (1983) 575–582.

20   Grigorescu, F., Flier, J. S., and Kahn, C. R., Defect in insulin receptor phosphorylation in erythrocytes and fibroblasts associated with severe insulin resistance. J. biol. Chem. *259* (1984) 15003–15006.

21   Grigorescu, F., White, M. F., and Kahn, C. R., Insulin binding and insulin-dependent phosphorylation of the insulin receptor solubilized from human erythrocytes. J. biol. Chem. *258* (1983) 13708–13716.

22   Grunberger, G., Comi, R. J., Taylor, S. I., and Gorden, P., Tyrosine kinase activity of the insulin receptor of patients with type A extreme insulin resistance; studies with circulating mononuclear cells and cultured lymphocytes. J. clin. Endocr. Metab. *59* (1984) 1152–1158.

23   Grunberger, G., Zick, Y., and Gorden, P., Defect in phosphorylation of insulin receptors in cells from an insulin-resistant patient with normal insulin-binding. Science *223* (1984) 932–934.

24   Häring, H. U., Kasuga, M., and Kahn, C. R., Insulin receptor phosphorylation in intact adipocytes and in a cell-free system. Biochem. biophys. Res. Commun. *108* (1982) 1538–1545.

25   Häring, H. U., Kasuga, M., White, M. F., Crettaz, M., and Kahn, C. R., Phosphorylation and dephosphorylation of the insulin receptor: evidence against an intrinsic phosphatase activity. Biochemistry *23* (1984) 3298–3306.

26   Häring, H. U., Machicao, F., Kirsch, D., Rinninger, F., Hölzl, J., Eckel, J., and Bachmann, W., Protein kinase activity of the insulin receptor from muscle. FEBS Lett. *176* (1984) 229–234.

27   Häring, H. U., White, M. F., Kahn, C. R., Kasuga, M., Lauris, W., Fleischmann, R., Murray, M., and Pawelek, J., Abnormality of insulin binding and receptor phosphorylation in an insulin-resistant melanoma cell line. J. Cell Biol. *99* (1984) 900–908.

28   Hedo, J. A., Kasuga, M., Van Obberghen, E., Roth, J., and Kahn, C. R., Direct demonstration of glycosylation of insulin receptor subunits by biosynthetic and external labelling: Evidence for heterogeneity. Proc. natn. Acad. Sci. USA *78* (1981) 4791–4795.

29   Heldin, C. H., and Westermark, B., Growth factors; Mechanism of action and relation to oncogenes. Cell *37* (1984) 9–20.

30   Hunter, T., Synthetic peptide substrates for a tyrosine protein kinase. J. biol. Chem. *257* (1982) 4843–4848.

31   Hunter, T., The proteins of oncogenes. Scient. Am. *251* (1984) 60–69.

32   Hunter, T., and Sefton, B. M., Transforming gene product of Rous Sarcoma virus phosphorylates tyrosine. Proc. natn. Acad. Sci. USA *77* (1980) 1311–1315.

33   Jacobs, S., and Cuatrecasas, P., Insulin: Structure and function. Endocr. Rev. *2* (1981) 251–263.

34   Jacobs, S., Hazum, E., Schechter, Y., and Cuatrecasas, P., Insulin receptor: covalent labeling and identification of subunits. Proc. natn. Acad. Sci. USA *76* (1979) 4918–4921.

35   Jacobs, S., Kull, F. C., Earp, H. S., Svoboda, M. E., VanWyk, J. J., and Cuatrecasas, P., Somato-medin-C stimulates the phosphorylation of the β-subunit of its own receptors. J. biol. Chem. *258* (1983) 9581–9584.

36   Jacobs, S., Schechter, Y., Bisell, K., and Cuatrecasas, P., Purification and properties of insulin receptor from rat liver membranes. Biochem. biophys. Res. Commun. *77* (1977) 981–988.

37   Kadowaki, T., Kasuga, M., Akanum, Y., Ezaki, D., and Takuku, F., Decreased autophosphory-lation of the insulin receptor-kinase in streptozotocin-diabetic rats. J. biol. Chem. *259* (1984) 14208–14216.

38   Kahn, C. R., Baird, K. L., Flier, J. S., Grunfeld, C., Harmon, J. T., Harrison, L. C., Karlsson, F. A., Kasuga, M., King, G. L., Lang, U. C., Podskalny, F. M., and Van Obberghen, E., Insulin receptors, receptor antibodies, and the mechanism of insulin action. Recent Prog. Horm. Res. *37* (1981) 447–538.

39   Kasuga, M., Karlsson, F. A., and Kahn, C. R., Insulin stimulates the phosphorylation of the 95,000-dalton subunit of its own receptor. Science *215* (1982) 185–187.

40   Kasuga, M., Fujita-Yamaguchi, Y., Blithe, D. L., and Kahn, C. R., Tyrosine-specific protein kinase activity is associated with the purified insulin receptor. Proc. natn. Acad. Sci. USA *80* (1983) 2137–2141.

41   Kasuga, M., Fujita-Yamaguchi, Y., Blithe, D. L., White, M. F., and Kahn, C. R., Character-ization of the insulin receptor kinase purified from human placental membranes. J. biol. Chem. *258* (1983) 10973–10980.

42   Kasuga, M., Zick, Y., Blithe, D. L., Crettaz, M., and Kahn, C. R., Insulin stimulates tyrosine phosphorylation of the insulin receptor in cell-free system. Nature *298* (1982) 667–669.

43   Kasuga, M., Zick, Y., Blithe, D. L., Karlsson, F. A., Häring, H. U., and Kahn, C. R., Insulin stimulation of phosphorylation of the beta subunit of the insulin receptor. Formation of both phosphoserine and phosphotyrosine. J. biol. Chem. *257* (1982) 9891–9894.

44   Kowalski, A., Gazzano, H., Fehlmann, M., and Van Obberghen, E., Dephosphorylation of the hepatic insulin receptor: absence of intrinsic phosphatase activity in purified receptors. Biochem. biophys. Res. Commun. *117* (1983) 885–893.

45   Kull, F. C. Jr, Jacobs, S., Su, Y.-F., Svoboda, M. E., VanWyk, J. J., and Cuatrecasas, P., Mono-clonal antibodies to receptors for insulin and somatomedin-C. J. biol. Chem. *258* (1983) 6561–6566.

46   Le Marchand-Brustel, Y., Grémeaux, T., Ballotti, R., and Van Obberghen, E., Insulin receptor tyrosine kinase is defective in skeletal muscle of insulin-resistant obese mice. Nature *315* (1985) 676–679.

47   Massagué, J., Pilch, P. F., and Czech, M. P., A unique proteolytic cleavage site on the β subunit of the insulin receptor. J. biol. Chem. *256* (1981) 3182–3190.

48   Nemenoff, R. A., Kwok, Y. C., Shulman, G. I., Blackshear, P. J., Osathanondh, R., and Avruch, J., Insulin-stimulated tyrosine protein kinase. Characterization and relation to the insulin recep-tor. J. biol. Chem. *259* (1984) 5058–5065.

49   Nickameyer, W. S., and Wang, L. H., Nucleotide sequence of avian sarcoma virus UR2 and comparison of its transforming gene with other members of the tyrosine protein kinase oncogene family. J. Virol. *53* (1985) 879–884.

50   Petruzzelli, L. M., Ganguly, S., Smith, C. J., Cobb, M. H., Rubin, C. S., and Rosen, O. M., Insulin activates a tyrosine-specific protein kinase in extracts of 3T3-L1 adipocytes and human placenta. Proc. natn. Acad. Sci. USA *79* (1982) 6792–6796.

51   Petruzzelli, L. M., Herrera, R., and Rosen, O. M., Insulin receptor is an insulin-dependent ty-rosine protein kinase: copurification of insulinbinding activity and protein kinase activity to homogeneity from human placenta. Proc. natn. Acad. Sci. USA *81* (1984) 3327–3331.

44

52 Pike, L.J., Kuenzel, E.A., Casnellie, J.E., and Krebs, E.G., A comparison of the insulin- and epidermal growth factor-stimulated protein kinases from human placenta. J. biol. Chem. *259* (1984) 9913–9921.

53 Pilch, P.F., and Czech, M.P., Interaction of cross-linking agents with the insulin effector system of isolated fat cells covalent linkage of ^{125}I-insulin to a plasma membrane receptor protein of 140,000 daltons. J. biol. Chem. *254* (1979) 3375–3381.

54 Rees-Jones, R.W., Hedo, J.A., Zick, Y., and Roth, J., Insulin-stimulated phosphorylation of the insulin receptor precursor. Biochem. biophys. Res. Commun. *116* (1983) 417–422.

55 Rees-Jones, R.W., Hendricks, S.A., Quarum, M., and Roth, J., The insulin receptor of rat brain is coupled to tyrosine kinase activity. J. biol. Chem. *259* (1984) 3470–3474.

56 Rees-Jones, R.W., and Taylor, S.I., An endogenous substrate for the insulin receptor-associated tyrosine kinase. J. biol. Chem. *260* (1985) 4461–4467.

57 Reynolds, F.H. Jr, Todaro, G.J., Fryling, C., and Stephenson, J.R., Human transforming growth factors induce tyrosine phosphorylation of EGF receptors. Nature *292* (1981) 259–262.

58 Rosen, O.M., Herrera, R., Olowe, Y., Petruzzelli, L.M., and Cobb, M.H., Phosphorylation activates the insulin receptor tyrosine protein kinase. Proc. natn. Acad. Sci USA *80* (1983) 3237–3240.

59 Roth, R.A., Cassell, D.J., Wrong, K.Y., Maddux, B.A., and Goldfine I.D., Monoclonal antibodies to the insulin receptor block binding and inhibit insulin action. Proc. natn. Acad. Sci. USA *79* (1982) 7312–7316.

60 Roth, R.A., and Cassell, D.J., Insulin receptor: evidence that it is a protein kinase. Science *219* (1983) 299–301.

61 Roth, R.A., Cassell, D.J., Maddux, D.A., and Goldfine, I.D., Regulation of insulin receptor kinase activity by insulin mimickers and an insulin antagonist. Biochem. biophys. Res. Commun. *115* (1983) 245–252.

62 Roth, R.A., Cassell, D.J., Morgan, D.D., Tatnell, M.A., Jones, R.H., Schuttler, A., and Brandenburg, D., Effects of covalently linked insulin dimers on receptor kinase activity and receptor down regulation. FEBS Lett. *70* (1984) 360–364.

63 Roth, R.A., Mesirow, M.L., and Cassell, D.J., Preferential degradation of the β-subunit of purified insulin receptor. Effect on insulin binding and protein kinase activities of the receptor. J. biol. Chem. *258* (1983) 14456–14460.

64 Rubin, J.B., Shia, M.A., and Pilch, P.F., Stimulation of tyrosine-specific phosphorylation in vitro by insulin-like growth factor I. Nature *305* (1983) 438–440.

65 Sadoul, J.L., Peyron, J.F., Ballotti, R., Debant, A., Fehlmann, M., and Van Obberghen, E., Identification of a cellular 110,000 Da protein substrate for the insulin receptor kinase. Biochem. J. *227* (1985) 887–892.

66 Shia, M.A., and Pilch, P.F., The β-subunit of the insulin receptor is an insulin-activated protein kinase. Biochemistry *22* (1983) 717–721.

67 Shia, M.A., Rubin, J.D, and Pilch, P.F., The insulin receptor protein kinase. Physicochemical requirements for activity. J. biol. Chem. *58* (1983) 14450–14455.

68 Siegel, T.W., Ganguly, S., Jacobs, S., Rosen, D.M., and Cuatrecasas, P., Purification and properties of the human placenta insulin receptor. J. biol. Chem. *256* (1981) 9266–9273.

69 Simpson, I.A., and Hedo, J.A., Insulin receptor phosphorylation may not be a prerequisite for acute insulin action. Science *223* (1984) 1301–1304.

70 Stadtmauer, L.A., and Rosen, D.W., Phosphorylation of exogenous substrates by the insulin receptor-associated protein kinase. J. biol. Chem. *258* (1983) 6682–6685.

71 Tamura, S., Brown, T.A., Dubler, R.E., and Larner, J., Insulin-like effect of vanadate on adipocyte glycogen synthase and on phosphorylation of 95,000 dalton subunit of insulin receptor. Biochem. Res. Commun. *113* (1983) 80–86.

72 Tamura, S., Fujita-Yamaguchi, Y., and Larner, J., Insulin-like effect of trypsin on the phosphorylation of rat adipocyte insulin receptor. J. biol. Chem. *258* (1983) 14749–14752.

73 Ullrich, A., Bell, J.R., Chen, E.Y., Herrera, R., Petruzelli, L.M., Dall, T.J., Gray, A., Coussens, L., Liao, Y.C., Tsubokawa, M., Mason, A., and Seeburg, P.H., Human insulin receptor and its relationship to the tyrosine kinase family of oncogenes. Nature *313* (1985) 756–761.

74 Ullrich, A., Coussens, K., Hayflick, J.S., Dall, T.J., Gray, A., Tam, A.W., Lee, J., Yarden, Y., Libermann, T.A., Schlessinger, J., Downward, J., Mayes, E.L.V., Whittle, N., Waterfield, M.D., and Seeburg, P.H., Human epidermal growth factor receptor cDNA sequence and aberrant expression of the amplified gene in A431 epidermoid carcinoma cells. Nature *309* (1984) 418–425.

75 Van Obberghen, E., The insulin receptor: its structure and function. Biochem. Pharmac. *33* (1984) 889–896.

76 Van Obberghen, E., Ballotti, R., Gazzano, H., Fehlmann, M., Rossi, B., Gammeltoft, S., Debant, A., Le Marchand-Brustel, Y., and Kowalski, A., The insulin receptor kinase. Biochimie *67* (1985) 1119–1123.

77 Van Obberghen, E., Kasuga, M., Le Cam, A., Hedo, J. A., Itin, A., and Harrison, L. C., Biosynthetic labelling of the insulin receptor: studies of subunits in cultured human IM-9 lymphocytes. Proc. natn. Acad. Sci. USA *78* (1981) 1052–1056.

78 Van Obberghen, E., and Kowalski, A., Phosphorylation of the hepatic insulin receptor; stimulating effects of insulin on intact cells and in a cell-free system. FEBS Lett. *143* (1982) 179–182.

79 Van Obberghen, E., Rossi, B., Kowalski, A., Gazzano, H., and Ponzio, G., Receptor-mediated phosphorylation of the hepatic insulin receptor; evidence that the Mr 95,000 receptor subunit is its own kinase. Proc. natn. Acad. Sci. USA *80* (1983) 945–949.

80 Velicelebi, G., and Aiyer, R. A., Identification of the alpha beta monomer of the adipocyte insulin receptor by insulin binding and autophosphorylation. Proc. natn. Acad. Sci. USA *24* (1984) 7693–7697.

81 White, M. F., Häring, H. U., Kasuga, M., and Kahn, C. R., Kinetic properties and sites of autophosphorylation of the partially purified insulin receptor from hepatoma cells. J. biol. Chem. *259* (1984) 255–264.

82 Yip, C. C., Yeung, C. W. T., and Moule, M. L., Photoaffinity labeling of insulin receptor of rat adipocyte plasma membrane. J. biol. Chem. *253* (1978) 1743–1745.

83 Yu, K. T., and Czech, M. P., Tyrosine phosphorylation of the insulin receptor β subunit activates the receptor-associated tyrosine kinase activity. J. biol. Chem. *259* (1984) 5277–5286.

84 Yu, K. T., Werth, D. K., Pastan, I. H., and Czech, M. P., *Src* kinase catalyzes the phosphorylation and activation of the insulin receptor kinase. J. biol. Chem. *260* (1985) 5838–5845.

85 Zick, Y., Grunberger, G., Podskalny, J. M., Moncada, V., Taylor, S. I., Gorden, P., and Roth, J., Insulin stimulates phosphorylation of serine residues in soluble insulin receptors. Biochem. biophys. Res. Commun. *116* (1983) 1129–1135.

86 Zick, Y., Kasuga, M., Kahn, C. R., and Roth, J., Characterization of insulin-mediated phosphorylation of the insulin receptor in a cell-free system. J. biol. Chem. *258* (1983) 75–80.

87 Zick, Y., Rees-Jones, R. W., Taylor, S. I., Gorden, P., and Roth, J., The role of antireceptor antibodies in stimulating phosphorylation of the insulin receptor. J. biol. Chem. *259* (1984) 4396–4400.

88 Zick, Y., Whittaker, J., and Roth, J., Insulin stimulated phosphorylation of its own receptor. Activation of a tyrosine-specific protein kinase that is tightly associated with the receptor. J. biol. Chem. *258* (1983) 3431–3434.

Internalization of polypeptide hormones and receptor recycling

J.-L. Carpentier, P. Gorden*, A. Robert and L. Orci

*Institute of Histology and Embryology, University of Geneva Medical Center, CH–1211 Geneva 4 (Switzerland), and * Diabetes Branch, National Institute of Arthritis, Diabetes, and Digestive and Kidney Diseases, National Institutes of Health, Bethesda (Maryland 20205, USA)*

Introduction

The insulin receptor is an integral plasma membrane glycoprotein of most cells. It consists of 2 subunits linked by disulfide bonds; the alpha and beta subunits of the receptor are synthesized by way of a single chain proreceptor which is cleaved and further processed, by the addition of complex carbohydrates, prior to insertion into the plasma membrane[49]. Recently a cDNA encoding the proreceptor has been cloned and the protein sequence of the receptor elucidated[29, 85]. The alpha subunit of the receptor is the major binding component and the beta subunit is a tyrosine specific protein kinase which itself can be autophosphorylated.

In a separate series of events the insulin receptor complex can be removed from the cell surface by receptor mediated endocytosis. In the present review we will focus on receptor mediated endocytosis of insulin, its receptor, and the functional consequence of these events.

Historical perspective

We now know that when ^{125}I-insulin, under physiological conditions, binds to cell surface receptors on a variety of cell types that the hormone-receptor complex is internalized by cells. Subsequently, a series of intracellular events can then separate the hormone from the receptor allowing the hormone and receptor to be processed independently. It is instructive, therefore, to remember how we arrived at this point.

Studies carried out in the late 1960s and early 1970s demonstrated that polypeptide hormones such as ACTH[63] and insulin[27, 37] could be labeled with ^{125}I in a biologically active form that would bind in a specific fashion to cells or subcellular constituents. Since these studies were carried out at low temperature to minimize hormone degradation, attention was focused on the cell surface and a fully reversible system.

Thus, in 1976 we started with the assumption that insulin bound to cell surface receptors in a fully reversible manner and that if we could develop an appropriate probe we could visualize this interaction directly at the electron micro-

48

scopic level. Several electron dense probes were available, but because of problems of specificity, we settled on EM autoradiography. Though it was necessary to develop quantitative means of analysis, this methodology could be used analogously with biochemical studies. Our initial results, incubating ^{125}I-insulin with IM-9 lymphocytes, confirmed what we assumed to be true, i.e., that the hormone bound to the cell surface. Shortly afterwards, however, in collaboration with Pierre Freychet and Alphonse LeCam, we found that when ^{125}I-insulin was incubated with rat hepatocytes at 37°C that the labeled material not only binds to the cell surface, but is extensively internalized by the cell[20]. The apparent discrepancy with the IM-9 studies was only a matter of degree since IM-9 cells do internalize insulin[13, 46], but at a slow rate and to a very small extent compared to hepatocytes[15, 17, 18].

Thus, by early 1977 it was clear that polypeptide hormones not only bound to cell surface receptors[20], but also shared post binding events with ligands as diverse as low density lipoproteins[1]. It soon became clear that many other polypeptide hormones and growth factors as well as other unrelated ligands such as $\alpha2$-macroglobulin[87], asialoglycoprotein[55], transferrin[56] and others shared many common features upon interacting with cells.

Because the events related to insulin appear to be common for other polypeptide hormones, growth factors and other transport receptor proteins, we will use these examples for a broader perspective[43, 44]. Finally, we will attempt to clarify certain areas of apparent controversy. The morphologic events that occur following hormone binding to receptor have been previously discussed[43, 44] and are schematized in figure 1.

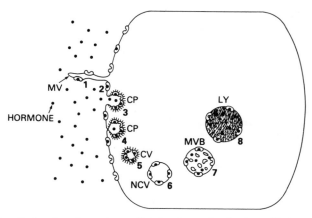

Figure 1. Schematic drawing of receptor mediated endocytosis of polypeptide hormones. Step 1 indicates initial preferential association to microvilli (MV); steps 2 and 3 represent redistribution of the hormone-receptor complex to undifferentiated plasma membrane and coated pits (CP); steps 4 and 5 represent the formation of the earliest endocytotic vesicle which is coated (CV); step 6 represents the presumed next step which involves a non-coated endocytotic vesicle (NCV); and steps 7 and 8 represent the evolution of the various lysosomal forms, i.e., multivesicular body (MCV) and various types of more dense lysosomes (LY). Reproduced from ref. 43 with permission.

Events at the cell surface (steps 1, 2 and 3)

There are 3 features of the cell surface, discernable morphologically, that have been studied in detail. For some cells that bind insulin, the surface is relatively smooth (fig. 2). For many cells, however, the surface is covered with numerous microvilli; this is true for cultured human lymphocytes and freshly isolated rat hepatocytes (fig. 3). In the cultured human lymphocyte, microvilli constitute ~ 55% of the cell surface. These structures provide a large surface for initial contact with the ligand. The villi are variable in length, contain abundant cytoskeletal structures, increased density of intramembrane particles, and are largely devoid of cytoplasmic organelles[23].

Small segments of the plasma membrane are decorated with a cytoplasmic bristle coat made up predominantly of the protein clathrin[73] and at least two other minor proteins[89]. These coated segments predominate in invaginated portions of the membrane and are referred to as coated pits[1]. These pits range in size from ~ 60 nm in lymphocytes[23] and hepatocytes[17] to ~ 200 nm in human fibroblasts[1,70] and 3T3-L1 adipocytes[31,33] (figs 2 and 4). Though they occur in essentially all cells, their importance with respect to receptor mediated endocytosis was first recognized by Anderson et al., who found that the internalization

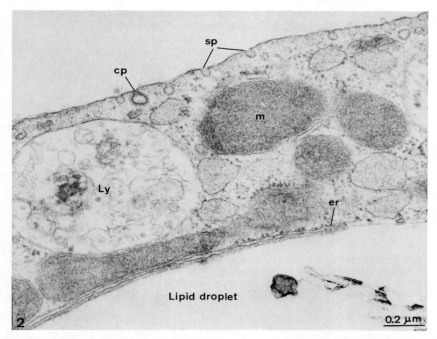

Figure 2. Low power view of the periphery of a 3T3-L1 adipocyte in culture showing a small portion of a lipid droplet, and various organelles, i.e., er, endoplasmic reticulum; m, mitochondria; ly, lysosome. The surface is relatively smooth (compare with fig. 5) but invaginates to form coated pits (cp) and small apparently uncoated invaginations (sp), × 55,000.

50

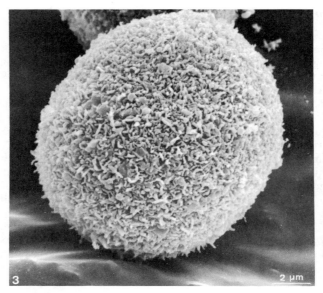

Figure 3. General view of a freshly isolated rat hepatocyte seen at the scanning microscope. The surface of the cell is filled with microvillosities and cytoplasmic projections, × 5,000.

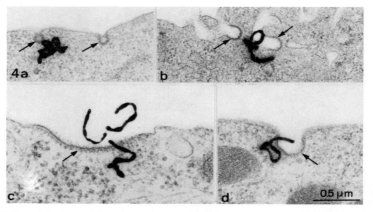

Figure 4. Coated pits (arrows) in various cell types. *a* IM-9 cultured human lymphocyte incubated with ^{125}I-insulin; *b* freshly isolated rat hepatocyte incubated with ^{125}I-monoiodo biosynthetic human insulin; *c* cultured human fibroblast incubated in the presence of ^{125}I-EGF and LDL-ferritin; *d* 3T3-L1 adipocyte incubated with ^{125}I-insulin. *a* × 23,0000; *b* × 23,000; *c* × 35,000; *d* × 25,000.

of low density lipoprotein occurred essentially exclusively by way of coated pits[1].

Coated pits are distinguished morphologically in several other ways: on freeze-etched preparations they have an increased number of intramembrane parti-

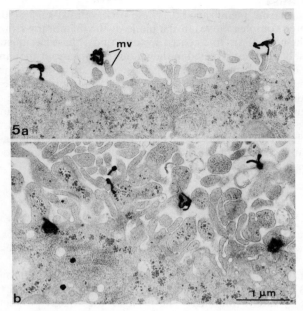

Figure 5. Periphery of freshly isolated rat hepatocytes incubated for short periods of time (2 min) at 20 °C in the presence of A-14 monoiodo biosynthetic human insulin. As seen both in section perpendicular to the plane of the plasma membrane (a) and in section tangential to this plane, autoradiographic grains are mostly associated with microvilli (mv) present on the cell surface. × 14,000.

cles[70]; on quick frozen deep etched preparations these structures have a basket shaped appearance comprising numerous pentagons and hexagons[53]. The cholesterol affinity probe, filipin, disrupts the undifferentiated portion of the plasma membrane while having little effect on coated segments[68]. The explanation for this is uncertain since filipin should have access to cholesterol on both the coated and uncoated segment and the cholesterol content is believed to be similar in both segments[81]. Regardless of the explanation, however, filipin provides a useful morphologic marker.

The remainder of the surface of most cells is comprised of a smooth trilaminar surface and apparently uncoated small invaginations. The role of the small invagination, which may be abundant in 3T3-L1 adipocytes or isolated rat adipocytes, is unclear[33].

When ^{125}I-insulin is incubated with either 3T3-L1 adipocytes, cultured human lymphocytes, or freshly isolated rat hepatocytes, autoradiographic grains localize initially and preferentially to the microvillous surface of the cell (fig. 5)[23, 31]. Preferential localization also occurs to coated pits[31] (fig. 4). With time there is a redistribution of the labeled material to the non-villous and coated portion of the membrane. This occurs under circumstances where the ligand continuously binds to receptor. Studies using both radioactively labeled[3] and fluorescently labeled ligands have shown that the ligand-receptor complex is mobile in the plane of the plasma membrane[76, 77]. These observations suggest that the hormone-receptor complex moves in the plane of the membrane from the villous surface to the non-villous and coated surface; under these circumstances receptors presumably are not repleted in the villous membrane at the same rate as in other parts of the plasma membrane. These data are also consistent with the idea that the affinity for insulin may be higher in receptors in the non-villous portion of the membrane. This point is emphasized by the finding of a correlation between slowed dissociation of ^{125}I-insulin from the plasma membrane of cells and the redistribution of the ligand on the surface of the cell. Thus, when ^{125}I-insulin binds to the cell surface at low temperature and is allowed to dissociate, an increasing proportion of the ligand is redistributed to the non-villous and coated surface of the cell[11].

Initial steps of endocytosis (steps 4 and 5)

While we have indicated that internalization may occur in non-coated areas of the membrane, the process is best understood by studies of the coated pit mechanism.

Coated pits are exposed to the external environment of the cell; thus the ligand either binds initially to receptors in these structures or the complex moves into coated pits by lateral mobility. The neck of the coated pit then fuses and subsequently fissions from the membrane surface to form a coated vesicle (fig. 6). These vesicles have been demonstrated directly morphologically by serial sections which have verified that these are closed vesicular structures[31, 51, 74, 86]. The lifetime of the coated vesicle is short, i.e., 1–2 min or less and by some unknown mechanism, the coat is shed from the membrane-limited vesicle. Recent studies suggest that clathrin in coated vesicles is initially acted upon by a 70,000 mol.wt polypeptide containing ATPase activity. This uncoating ATPase presumably acts prior to clathrin release. It is of special note that this uncoating ATPase greatly prefers closed cages to free clathrin triskelions as substrates. Thus, coated vesicles are preferred to coated pits[8, 78, 79]. An alternative thesis has been proposed by Willingham and Pastan, who suggest that the coated membrane never separates from the cell surface, but instead the initial vesicle formed is uncoated[72, 88]. This controversy awaits further study for clarification.

Later steps of endocytosis (steps 6, 7 and 8)

The coated vesicle, as mentioned, has a short half-life and the next structure with which the labeled ligand associates is a larger clear vesicle or endosome[32] (fig. 7). These structures have been shown to have important functional properties. Maxfield and associates have shown that the endosome has a proton pump in its membrane capable of acidifying the internal milieu of the vesicle[84]. This acidic environment has important functional consequences for many different types of ligands: for viruses, acidification allows the viral membrane to fuse with the vesicle membrane so that the virus can penetrate the vesicular mem-

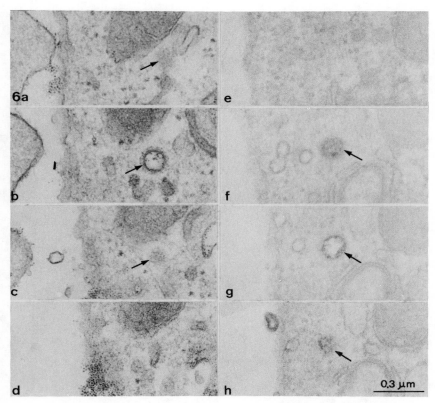

Figure 6. Serial sections through coated vesicles that have internalized cationic ferritin (a–d) or horseradish peroxidase (HRP) (e–f). As the sections progress from top to bottom, the vesicles (arrows) are identified in various planes. The examples shown here have no surface connection and are true vesicles. In the case of HRP incubation the fixed cells were stained en bloc and the sections not counterstained. × 45,000.

brane and gain access to the cytoplasm[64]; for the transferrin system, acidification allows diferric iron to separate from its carrier protein and gain access to the cell[62]; for many other ligands, including the polypeptide hormones, acidification promotes dissociation of ligand from receptor, thus allowing the ligand to be sequestered and processed separately from the receptor.

The next structure seen is the multivesicular body (fig. 7). This organelle is larger than the clear endosome and is further characterized by occasional or many small vesicles. The biogenesis of these structures is unclear; whether they exist de novo or are formed by the fusion of endosomes and other small vesicles remains to be established. The multivesicular body appears to be an intermediate in a continuous process; the next structure visualized is, therefore, the lysosome[32] (fig. 7). Originally the lysosome was thought to be unique in two ways: a) by containing a proton pump capable of rendering its internal environ-

54

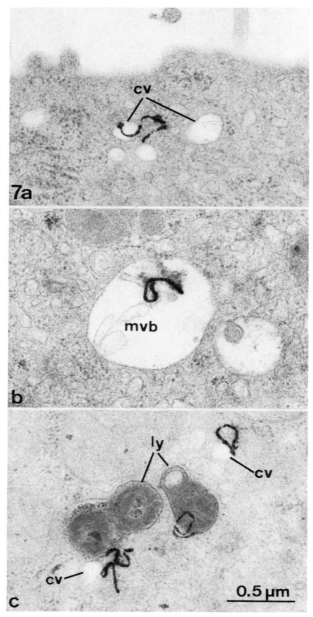

Figure 7. Thin section of isolated rat hepatocytes incubated for 60 min at 37°C with ^{125}I-insulin. Autoradiographic grains are seen overlying or in the vicinity of clear vesicles (a and c), multivesicular bodies (b) and dense bodies (c). (cv, clear vesicles; mvb, multivesicular bodies; ly, lysosomes). × 36,000. Reproduced from ref. 75 with permission.

ment acidic and b) by containing biologically active enzymes with an acidic pH optimum. This structure has, therefore, lost some of its novelty since the internal environment of both the endosome and multivesicular body are also acidic. Though the lysosome has distinctive morphologic features, strictly speaking it can be identified only by cytochemical techniques that reveal biologically active enzymes. These techniques may be relatively insensitive unless rigorously applied or multiple enzymes are sought. Thus, it is most appropriate to call structures like multivesicular bodies which have been considered a form of lysosome as 'lysosome-like.' Finally, ambiguity of nomenclature is avoided if one considers the endosome, multivesicular body and lysosome as a continuum involved in processing of ligands, their receptors or both.

Receptor recycling

The concept of recycling of plasma membrane components was first elucidated on kinetic grounds. Steinman and associates observed that the rapid turnover rate of the plasma membrane of L-cells required the de novo recycling of membrane components and was more rapid than could be explained by new protein synthesis[80, 82]. It was further shown, using a novel membrane iodination technique, that internal membranes such as lysosomal membranes could cycle to the plasma membrane[67]. Other studies indicated that exocytosis was coupled to endocytosis in such a way as to maintain cell shape and constancy of surface area[71].

For the polypeptide hormones such as insulin it has been shown both biochemically[50, 65] and morphologically[35, 45] that the receptor is internalized. It has further been shown morphologically that the receptor may be recycled to the plasma membrane[12, 36]. Thus, the receptor may be recycled, degraded or both situations may apply. The fate of the receptor following internalization may be the same or different from the ligand. That is, an uncoupling step may take place following initial endocytosis. The nature of this uncoupling step is obscure with respect to the polypeptide hormones.

In other systems, however, where large numbers of receptors are available for analysis, it has been shown that specific structures participate in the uncoupling step. For instance, when the asialoglycoprotein receptor and ligand were separately revealed using immunocytochemical techniques that involved colloidal gold particles of different size, separation of the two components was visualized morphologically[39]. The ligand appeared destined for the lysosome, whereas the receptor entered a tubulovesicular compartment (CURL) thought to be involved in returning the receptor to the plasma membrane.

Several features of the uncoupling process appear to be important. The early endocytotic vesicles has an acidic pH (pH $\sim$ 5.4) which promotes separation of the ligand and receptor[84]. It is well known that insulin is rapidly dissociated from its receptor under acidic conditions. While separation of ligand and receptor is not required for recycling, since insulin covalently bound to its receptor will recycle, it is likely that rapid recycling requires this separation. Thus, the acidic endosome is likely the first step in recycling. Morphologically speaking covalently bound insulin receptor complexes recycle in the same types of vesicles as described for endocytosis per se[12].

56

In other model systems it has recently been shown that an actively recycling receptor such as the transferrin receptor is further segregated into a compartment with a pH of 6.5, i.e., mildly acidic structures, whereas α-2 macroglobulin under similar conditions is transferred to a vesicle of pH 5.4[90]. It is further known that agents that inhibit acidification of endosomes, lysosomes and related structures also inhibit recycling[5, 10].

The precise morphologic steps involved in the recycling of the insulin receptor are unknown and we tentatively illustrate this as a continuous process of membrane retrieval (fig. 8). Whether the recycled vesicle enters the insertion pathway of de novo receptor synthesis is also unknown.

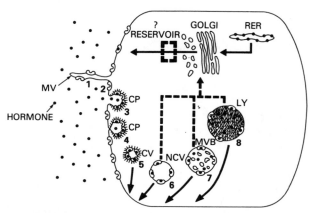

Figure 8. Synthesis and retrieval of polypeptide hormone receptors. Steps 1–8 are the same as shown in figure 1. The solid arrows represent points at which membrane receptors can recycle back to the plasma membrane. The dashed arrow indicates that recycling could also occur by way of the synthetic pathway for membrane integral proteins. A potential reservoir is shown to indicate, under some circumstances, there may be a pre-formed pool of intracellular receptors. Reproduced from ref. 43 with permission.

What determines the specificity of receptor mediated endocytosis?

We have indicated the general traffic pattern involved in the receptor mediated endocytosis of a polypeptide hormone. The structures involved appear to be common for a wide variety of ligands. For instance, cationic ferritin, a ligand that binds simply to anionic sites on the plasma membrane, is endocytosed by the same general structures as is insulin[31, 32, 34]. Further, we have shown that LDL and epidermal growth factor (EGF), two ligands that bind to different receptors, are taken up not only by the same type of structure, but by exactly the same structures. Thus, EGF and LDL co-localize to the same endosomes and lysosome-like structures during the process of endocytosis[14].

Thus, specificity lies in the receptor per se. The subsequent anatomical compartments are relatively non-specific. Another kind of specificity must reside at the level of recycling since receptors in the apparent same compartments are recycled at very different rates.

'Apparent' areas of controversy involving the intracellular localization of polypeptide hormones

In addressing areas of apparent controversy, it is important to emphasize that we are expressing a point of view not necessarily a resolution of the controversy. Up until this point we have emphasized the coated vesicle, endosome and lysosome-like structures in the intracellular movement of endocytosed ligands and their respective receptors (fig. 1). Other workers, however, have emphasized the nucleus and the Golgi as sites of intracellular localization.

Goldfine et al. have emphasized a direct nuclear action of insulin. The main point of concern here is whether ^{125}I-insulin can be localized to the nucleus of the cell morphologically. Two cell types have been studied, the IM-9 cultured human lymphocyte and the freshly isolated rat hepatocyte or perfused liver. We consistently find that the IM-9 cell internalizes insulin but to a very small extent[13, 46], whereas Goldfine et al. find more extensive internalization[41]. Since the methods of analysis are similar, the reasons for these differences are unclear. Using a photoreactive insulin, Olefsky also finds a very low internalization rate in IM-9 cells[7]. While all agree that ^{125}I-insulin is not localized to the nucleus per se in either hepatocytes or IM-9 cells[17, 40, 41, 46], Goldfine et al. find localization to the nuclear membrane of these cells[40, 41] and we do not. Thus, the controversy here is technical.

Posner, Bergeron and their associates have emphasized the initial localization of ^{125}I-insulin to the Golgi[6, 59]. Golgi localization has been based on two types of studies. When ^{125}I-insulin was injected into the rat and allowed to circulate for a few minutes, and the liver tissue removed from the sacrificed rat was homogenized and fractionated by isopycnic techniques, it was shown that labeled insulin co-migrates with Golgi marker enzymes. The second type of study was morphologic where it has been suggested that labeled insulin localizes with high frequency to lipoprotein filled vesicles. These vesicles are considered to be Golgi related.

In our initial studies using EM autoradiography in both hepatocytes and lymphocytes we found labeled insulin preferentially localized to lysosome-like structures. Though we indicated that some form of endocytotic vesicle must be interposed, we were unable to specifically identify these structures due to their small size[13]. Using 3T3-L1 adipocytes, however, an endosomal structure could be clearly identified[32]. More recently, in experiments employing colchicine which slows the passage of ligand from the plasma membrane to the lysosome, we have clearly identified an endosomal compartment in rat hepatocytes[75]. Further, neither we nor others have ever localized labeled insulin to the Golgi cisternae[19, 32]. Thus, the issue is whether the ligand is localized in lipoprotein-rich vesicles distal to the 'trans' Golgi cisternae. These vesicles, however, may be either secretory or endocytotic and are not specific for Golgi[25]. It is possible that more specific approaches will be able to distinguish the secretory from endocytotic nature of these vesicles[52].

Thus, we see this controversy as being in part semantic. In our view, labeled insulin is initially internalized into an endosomal compartment. These endosomes appear to have similar but not identical isopycnic properties to the structures containing Golgi markers and, therefore, the term Golgi fractions does not seem appropriate to describe these vesicular structures.

Biochemical features of receptor mediated endocytosis

The plasma membrane of cells turns over by endocytosis on a continuous basis. Macromolecules, therefore, are continuously taken up by adsorptive or receptor mediated endocytosis. Since binding concentrates the ligand, adsorptive endocytosis is a much more efficient process. Adsorptive endocytosis may proceed continuously, be ligand induced or both. For instance, agents that inhibit receptor recycling may augment cell surface receptor loss for LDL[5]. This suggests that, at least in part, receptors are internalized and recycled on a continuous basis. On the other hand, recycling inhibitors do not alter the insulin receptor over short periods of incubation unless the ligand is present[10]. This suggests that endocytosis is ligand induced.

The mechanism by which the ligand induces endocytosis, however, is unknown. It was initially suggested on the basis of inhibitor studies that transglutaminases were involved[28]. This thesis was based on the idea that several chemical agents inhibited both transglutaminase activity and endocytosis. It is now clear, however, that the various amines and other agents that may affect enzymatic activity of transglutaminases do not inhibit endocytosis[47, 60].

More recently it has been shown that the receptors for EGF[26], insulin[58] and PDGF[30] undergo a specific ligand-induced tyrosine-specific autophosphorylation. Since these three ligands and their receptors are taken up by cells in a similar way[16, 42, 69], it is tempting to speculate that receptor phosphorylation might represent a general mechanism that triggers endocytosis. This would be consistent with the idea that this is an energy consuming process. Recently it has been suggested that phosphorylation may be an important mechanism by which the endocytosis of the transferrin receptor is triggered[61, 66]. This is based on phorbol ester induced phosphorylation of the receptor and in this case serine phosphorylation is implicated.

We have recently carried out a series of experiments to test the hypothesis that phosphorylation might be a general mechanism involved in receptor mediated endocytosis of polypeptide hormones. Insulin is internalized by IM-9 lymphocytes, and its receptor phosphorylation is also triggered by insulin binding. We reasoned that if phosphorylation were a general mechanism, the growth hormone receptor of these cells should also be phosphorylated. However, using a variety of conditions and the insulin receptor as a positive control, we have been unable to demonstrate autophosphorylation of the growth hormone receptor[2]. Finally, we have shown that an autoantibody directed against the insulin receptor and manifesting insulinomimetic properties is internalized by cells in an analogous fashion to insulin[24, 57]. This autoantibody does not stimulate receptor autophosphorylation[92]. Thus, our current concept is that phosphorylation is not the general mechanism triggering polypeptide hormone endocytosis.

Functional implications of receptor mediated endocytosis

The functional implications of receptor mediated endocytosis depend totally on the ligand. For instance, the LDL receptor mediates the internalization of the cholesterol-containing lipoprotein; cholesterol is liberated by lysosomal enzyme activity and, in turn, the free cholesterol regulates several key enzymes as well as the LDL receptor. Viruses gain entry into the cell and the acid milieu of the endosome allows for fusion of the viral membrane with the vesicle membrane resulting in escape of the virus into the cytoplasm. For transferrin, the endocytotic process regulates the delivery of iron to the cell. For the polypeptide hormones and growth factors, however, the main functional consequences of receptor mediated endocytosis appear to be hormone degradation and receptor regulation.

Different polypeptide hormones transduce an intracellular signal in different ways following their binding to cell surface receptors. Some polypeptide hormones, such as glucagon, activate adenylate cyclase by a complex coupling mechanism. Others, such as insulin, EGF and other growth factors work through an as yet obscure mechanism possibly in part involving receptor autophosphorylation. Yet all of these and other polypeptide hormones undergo a very similar pattern of endocytosis. It is possible that intracellular transport of phosphoproteins subserves some function, but the nature of this putative function is unknown. Thus, the possible role of endocytosis in hormone action, if any, is obscure.

If cell surface binding activates hormone action, then there must be a regulatory mechanism to terminate that action. For instance, insulin degradation has been shown to be receptor linked[83]. Thus, internalization may be an important mechanism for removing the hormone from the cell surface and initiating its degradation. For certain polypeptides such as EGF and growth hormone, it is likely that ligand degradation is primarily lysosomal. This is supported by the lysosomal localization of these peptides and inhibition of their degradation by agents that alkalinize the endosome and lysosome. A similar process is true for insulin in that insulin is internalized into exactly the same compartments as are the other hormones, and insulin degradation is also inhibited by so-called lysosomotropic agents such as chloroquine, ammonium chloride and other similar compounds[47]. In the case of insulin, however, neutral proteases also have been implicated in its degradation[47]. While there is no question that soluble neutral proteases will rapidly degrade insulin, it is not totally clear that these proteases have access to the hormone in the intact cell. It must be remembered that the early endocytotic vesicles are also acidic, making it unlikely that these neutral proteases are active during the early phases of endocytosis.

Our current thinking is that insulin degradation occurs in part through a lysosomal mechanism but that neutral proteases may also play a role. Whether these proteases act solely on the cell surface, however, remains to be determined.

Another general feature or consequence of hormone binding to receptor is ligand induced receptor regulation[38]. Since both the hormone and receptor are internalized it is clear that endocytosis is an important mechanism of modulating cell surface receptor concentration. Further, it is clear that there may be no simple relationship between ligand and receptor internalization and receptor down regulation.

This is true because several regulatory events are involved. Growth hormone in cultured lymphocytes, and EGF binding in human fibroblasts, are two relatively straightforward examples. For instance, small concentrations of hormone induced receptor down regulation, and receptor down regulation and ligand internalization, are highly correlated as a function of time[4, 9, 42, 54]. In the case of growth hormone the ligand is, in part, irreversibly bound, and down regulation is further enhanced by ammonium chloride[4]. Thus, in this system there appears to be little recycling and what recycling does occur is inhibited by alkanization and inhibition of ligand dissociation.

The lack of a straightforward relationship between down regulation and internalization relates to recycling. We have pointed out that at some step in the endocytotic pathway the ligand and the receptor may become uncoupled and each may have a different fate. Thus, the receptor may be recycled, may be degraded, or some combination of these events may take place. In the U-937 human monocyte cell line, small concentrations of insulin induce the loss of cell surface receptors. These receptors are in part recovered in soluble extracts of the cell. Thus, in this case, receptors appear to be internalized and in part recovered intact intracellularly and in part are degraded. Monensin, which inhibits recycling, augments down regulation. Monensin, however, does not affect that fraction of receptors destined for degradation. Thus, monensin affects only the recycling component of the receptor. In other cell types where the rate of internalization is low, such as in the IM-9 lymphocytes, monensin has essentially no effect[10].

Hormone and receptor internalization also provide a mechanism for understanding receptor regulation in clinical states. Thus, nontarget cells like peripheral blood monocytes have insulin receptors. The monocyte insulin receptor is regulated in an analogous fashion to hepatocyte receptors. Monocytes and hepatocytes exposed to the same concentration of insulin would, therefore, regulate their receptors in an analogous fashion by way of receptor mediated endocytosis[48].

Further evidence that receptor mediated endocytosis is a regulated process comes from studies in hypoinsulinemic states. If it is true that endocytosis is a mechanism to remove the ligand from the cell surface and terminate its signal, it might be expected that in hypoinsulinemic states that this process would be impaired. We have recently found that in streptozotocin induced diabetes in the rat, a hypoinsulinemic state, the internalization of ^{125}I-insulin is impaired. By contrast, the internalization of ^{125}I-glucagon is either normal or increased[22]. Further in insulinopenic type I diabetes of man, ^{125}I-insulin internalization is markedly inhibited[21].

Conclusion

The insulin receptor is an integral protein of the plasma membrane of the cell. It is composed of two subunits: an α subunit, which binds the hormone, and a β subunit which is a tyrosine specific protein kinase capable of undergoing autophosphorylation. These independent subunits are synthesized by way of a higher molecular weight single chain precursor and thus are the product of a single gene[29, 49, 85] localized to chromosome 19[29, 91]. Assuming that the insulin receptor is synthesized in the same fashion as other integral membrane glycoproteins, then the nucleus, the rough endoplasmic reticulum, and the Golgi apparatus are involved in its biosynthesis. Further, there must be some form of transport of the mature receptor subunits to the plasma membrane where they are inserted.

By contrast, the endocytotic route involves coated pits, coated vesicles, large clear vesicles or endosomes, multivesicular bodies and other lysosomal forms. In addition, it is possible that some other as yet unidentified organelle is involved in recycling (fig. 8). At the present time, with respect to the insulin receptor, the biosynthetic pathway and the endocytotic pathway appear to be separate. Further, it does not appear that either pathway, i.e. synthesis or endocytosis, exerts a regulatory function over the other.

Acknowledgment. These studies were supported by Grant 3.460.83 of the Swiss National Science Foundation.

1   Anderson, R. G. W., Brown, M. S., and Goldstein, J. L., Role of the coated endocytic vesicle in the uptake of receptor-bound low density lipoprotein in human fibroblasts. Cell *10* (1977) 351–364.
2   Asakawa, K., Grunberger, G., McElduff, A., and Gorden, P., Polypeptide hormone receptor phosphorylation: is there a role in receptor mediated endocytosis? Endocrinology *117* (1985) 631–637.
3   Barazzone, P., Carpentier, J.-L., Gorden, P., Van Obberghen, E., and Orci, L., Polar redistribution of ^{125}I-labelled insulin on the plasma membrane of cultured human lymphocytes. Nature *286* (1980) 401–403.
4   Barazzone, P., Lesniak, M. A., Gorden, P., Van Obberghen, E., Carpentier, J.-L., and Orci, L., Binding, internalization, and lysosomal association of ^{125}I-human growth hormone in cultured human lymphocytes: a quantitative morphological and biochemical study. J. Cell Biol. *87* (1980) 360–369.
5   Basu, S. K., Goldstein, J. L., Anderson, R. G. W., and Brown, M. S., Monensin interrupts the recycling of low density lipoprotein receptors in human fibroblasts. Cell *24* (1981) 493–502.
6   Bergeron, J. J. M., Sikstrom, R., Hand, A. R., and Posner, B. I., Binding and uptake of ^{125}I-insulin into rat liver hepatocytes and endothelium. An in vivo radioautographic study. J. Cell Biol. *80* (1979) 427–443.
7   Berhanu, P., and Olefsky, J. M., Photoaffinity labeling of insulin receptors in viable cultured human lymphocytes: demonstration of receptor shedding and degradation. Diabetes *31* (1982) 410–417.
8   Braell, W. A., Schlossman, D. M., Schmid, S. L., and Rothman, J. E., Dissociation of clathrin coats coupled to the hydrolysis of ATP: Role of an uncoating ATPase. J. Cell Biol. *99* (1984) 734–741.
9   Carpenter, G., and Cohen, S., ^{125}I-labeled human epidermal growth factor binding, internalization and degradation in human fibroblasts. J. Cell Biol. *71* (1976) 159–171.
10 Carpentier, J.-L., Dayer, J.-M., Lang, U., Silverman, R., Orci, L., and Gorden, P., Down regulation and recycling of insulin receptors: effect of monensin on IM-9 lymphocytes and U-937 monocyte-type cells. J. biol. Chem. *259* (1984) 14 180–14 195.
11 Carpentier, J.-L., Fehlmann, M., Van Obberghen, E., Gorden, P., and Orci, L., Redistribution of ^{125}I-insulin on the surface of rat hepatocytes as a function of dissociation time. Diabetes *24* (1985) 1002–1007.
12 Carpentier, J.-L., Fehlmann, M., Van Obberghen, E., Gorden, P., and Orci, L., Morphological aspects and physiological relevance of insulin receptor internalization and recycling, in: Proceedings of the VII International Congress of Endocrinology, pp. 349–353. Eds F. Labrie and L. Proulx. Elsevier Science Publications, Amsterdam 1984.
13 Carpentier, J.-L., Gorden, P., Amherdt, M., Van Obberghen, E., Kahn, C. R., and Orci, L., ^{125}I-insulin binding to cultured human lymphocytes. Initial localization and fate of hormone determined by quantitative electron microscope autoradiography. J. clin. Invest. *61* (1978) 1057–1070.
14 Carpentier, J.-L., Gorden, P., Anderson, R. G. W., Goldstein, J. L., Brown, M. S., Cohen, S., and Orci, L., Co-localization of ^{125}I-epidermal growth factor and ferritin-low density lipoprotein in coated pits: a quantitative electron microscopic study in normal and mutant human fibroblasts. J. Cell Biol. *95* (1982) 73–77.

62

15 Carpentier, J.-L., Gorden, P., Barazzone, P., Freychet, P., LeCam, A., and Orci, L., Intracellular localization of ^{125}I-labeled insulin in hepatocytes from intact rat liver. Proc. natn. Acad. Sci. USA 76 (1979) 2803–2807.

16 Carpentier, J.-L., Gorden, P., Freychet, P., Canivet, B., and Orci, L., The fate of ^{125}I-iodoepidermal growth factor in isolated hepatocytes: A quantitative electron microscopic autoradiographic study. Endocrinology 109 (1981) 768–775.

17 Carpentier, J.-L., Gorden, P., Freychet, P., LeCam, A., and Orci, L., Lysosomal association of internalized ^{125}I-insulin in isolated rat hepatocytes. Direct demonstration by quantitative electron microscopic autoradiography. J. clin. Invest. 63 (1979) 1249–1261.

18 Carpentier, J.-L., Gorden, P., Freychet, P., LeCam. A., and Orci, L., Relationship of binding to internalization of ^{125}I-insulin in isolated rat hepatocytes. Diabetologia 17 (1979) 379–384.

19 Carpentier, J.-L., Gorden, P., Freychet, P., LeCam, A., and Orci, L., ^{125}I-insulin is preferentially internalized in regions of the hepatocytes rich in lysosomal and Golgi structures. Experientia 35 (1979) 904–905.

20 Carpentier, J.-L., Gorden, P., LeCam, A., Freychet, P., and Orci, L., Limited intracellular translocation of ^{125}I-insulin in isolated rat hepatocytes. Diabetologia 13 (1977) 386 (abstract).

21 Carpentier, J.-L., Grunberger, G., Robert, A., Orci, L., and Gorden, P., Regulation of receptor-mediated endocytosis of ^{125}I-insulin in hypoinsulinemic states: differential response in insulin dependent Type I diabetes vs. non-insulin dependent diabetes. Diabetes 34 Suppl. 1 7A, 1985 (abstract).

22 Carpentier, J.-L., Robert, A., Van Obberghen, E., Freychet, P., Gorden, P., and Orci, L., Inhibition of receptor-mediated endocytosis of ^{125}I-insulin into isolated hepatocytes of streptozotocin-treated diabetic rats: evidence for a regulated step in hypoinsulinemia. Program and Abstracts of the Endocrine Society. 22 (1985) (abstract).

23 Carpentier, J.-L., Van Obberghen, E., Gorden, P., and Orci, L., Surface redistribution of ^{125}I-insulin in cultured human lymphocytes. J. Cell Biol. 91 (1981) 17–25.

24 Carpentier, J.-L., Van Obberghen, E., Gorden, P., and Orci, L., Binding, membrane redistribution, internalization and lysosomal association of [^{125}I]anti-insulin receptor antibody in IM-9-cultured human lymphocyte. A comparison with [^{125}I]insulin. Expl Cell Res. 134 (1981) 81–92.

25 Chao, Y. S., Jones, A. L., Hradek, G. T., Windler, E. E. T., and Havel, R. J., Autoradiographic localization of the sites of uptake, cellular transport, and catabolism of low density lipoproteins in the liver of normal and estrogen-treated rats. Proc. natn. Acad. Sci. USA 78 (1981) 597–601.

26 Cohen, S., Carpenter, G., and King, L. Jr, Epidermal growth factor-receptor-protein kinase interactions: co-purification of receptor and epidermal growth factor-enhanced phosphorylation activity. J. biol. Chem. 255 (1980) 4834–4842.

27 Cuatrecasas, P., Insulin-receptor interactions in adipose tissue cells: direct measurement and properties. Proc. natn. Acad. Sci. USA 68 (1971) 1264–1268.

28 Davies, P. J. A., Davies, D. R., Levitzki, A., Maxfield F. R., Milhaud, P., Willingham, M. C., and Pastan, I. H., Transglutaminase is essential in receptor-mediated endocytosis of α_2-macroglobulin and polypeptide hormones. Nature 283 (1980) 162–167.

29 Ebina, Y., Ellis, L., Jarnagin, K., Edery, M., Graf, L., Clauser, E., Ou, J.-H., Masiarz, F., Kan, Y. W., Goldfine, I. D., Roth, R. A., and Rutter, W. J., The human insulin receptor cDNA: the structural basis for hormone-activated transmembrane signalling. Cell 40 (1985) 747–758.

30 Ek, B., Westermark, B., Wasteson, A., and Heldin, C.-H., Stimulation of tyrosine-specific phosphorylation by platelet-derived growth factor. Nature 295 (1982) 419–420.

31 Fan, J.-Y., Carpentier, J.-L., Gorden, P., Van Obberghen, E., Blackett, N. M., Grunfeld, C., and Orci, L., Receptor-mediated endocytosis of insulin: role of microvilli, coated pits, and coated vesicles. Proc. natn. Acad. Sci. USA 79 (1982) 7788–7791.

32 Fan, J.-Y., Carpentier, J.-L., Van Obberghen, E., Blackett, N. M., Grunfeld, C., Gorden, P., and Orci, L., The interaction of ^{125}I-insulin with cultured 3T3-L1 adipocytes: quantitative analysis by the hypothetical grain method. J. Histochem. Cytochem. 31 (1983) 859–870.

33 Fan, J.-Y., Carpentier, J.-L., Van Obberghen, E., Grunfeld, C., Gorden, P., and Orci, L., Morphological changes of the 3T3-L1 fibroblast plasma membrane upon differentiation to the adipocyte form. J. Cell Sci. 61 (1983) 219–230.

34 Fan, J.-Y., Carpentier, J.-L., Van Obberghen, E., Grunfeld, C., Gorden, P., and Orci, L., Surface interactions and intracellular localization of cationic ferritin: similarity to ^{125}I-insulin in 3T3-L1 adipocytes. Eur. J. Cell Biol. 31 (1983) 125–131.

35 Fehlmann, M., Carpentier, J.-L., LeCam, A., Thamm, P., Saunders, D., Brandenburg, D., Orci, L., and Freychet, P., Biochemical and morphological evidence that the insulin receptor is internalized with insulin in hepatocytes. J. Cell Biol. 93 (1982) 82–87.

36 Fehlmann, M., Carpentier, J.-L., Van Obberghen, E., Freychet, P., Thamm, P., Saunders, D., Brandenburg, D., and Orci, L., Internalized insulin receptors are recycled to the cell surface in rat hepatocytes. Proc. natn. Acad. Sci. USA 79 (1982) 5921–5925.

37 Freychet, P., Roth, J., and Neville, D. M. Jr, Insulin receptors in the liver: specific binding of [125I]insulin to the plasma membrane and its relation to insulin bioactivity. Proc. natn. Acad. Sci. USA 68 (1971) 1833–1837.

38 Gavin, J. R. III, Roth, J., Neville, D. M. Jr, De Meyts, P., and Buell, D. N., Insulin dependent regulation of insulin receptor concentrations. A direct demonstration in cell culture. Proc. natn. Acad. Sci. USA 71 (1974) 84–88.

39 Geuze, H. J., Slot, J. W., Strous, G. J. A. M., Lodish, H. F., and Schwartz, A. L., Intracellular site of asialoglycoprotein receptor-ligand uncoupling: double-label immunoelectron microscopy during receptor-mediated endocytosis. Cell 32 (1983) 277–287.

40 Goldfine, I. D., Jones, A. L., Hradek, G. T., and Wong, K. Y., Electron microscope autoradiographic analysis of [125I]iodoinsulin entry into adult rat hepatocytes in vivo: evidence for multiple sites of hormone localization. Endocrinology 108 (1981) 1821–1828.

41 Goldfine, I. D., Jones, A. L., Hradek, G. T., Wong, K. Y., and Mooney, J. S., Entry of insulin into human cultured lymphocytes; electron microscope autoradiographic analysis. Science, Wash. DC 202 (1978) 760–763.

42 Gorden, P., Carpentier, J.-L., Cohen, S., and Orci, L., Epidermal growth factor: morphological demonstration of binding, internalization, and lysosomal association in human fibroblasts. Proc. natn. Acad. Sci. USA 75 (1978) 5025–5029.

43 Gorden, P., Carpentier, J.-L., Fan, J.-Y., and Orci, L., Receptor mediated endocytosis of polypeptide hormones: mechanism and significance. Metabolism 31 (1982) 664–669.

44 Gorden, P., Carpentier, J.-L., Freychet, P., and Orci, L., Internalization of polypeptide hormones. Mechanism, intracellular localization and significance. Diabetologia 18 (1980) 263–274.

45 Gorden, P., Carpentier, J.-L., Moule, M. L., Yip, C. C., and Orci, L., Direct demonstration of insulin receptor internalization. A quantitative electron microscopic study of covalently bound 125I-photoreactive insulin incubated with isolated hepatocytes. Diabetes 31 (1982) 659–662.

46 Gorden, P., Carpentier, J.-L., Van Obberghen, E., Barazzone, P., Roth, J., and Orci, L., Insulin-induced receptor loss in the cultured human lymphocyte: quantitative morphological perturbations in the cell and plasma membrane. J. Cell Sci. 39 (1979) 77–78.

47 Gorden, P., Freychet, P., Carpentier, J.-L., Canivet, B., and Orci, L., Receptor linked degradation of 125I-insulin is mediated by internalization in isolated rat hepatocytes. Yale J. biol. Med. 55 (1982) 101–112.

48 Grunberger, G., Robert, A., Carpentier, J.-L., Dayer, J.-M., Roth, A., Stevenson, H. C., Orci, L., and Gorden, P., Human circulating monocytes internalize 125I-insulin in a similar fashion to rat hepatocytes: relevance to receptor regulation in target and non-target tissues. J. Lab. clin. Med. 106 (1985) 211–217.

49 Hedo, J. A., Kahn, C. R., Hayashi, M., Yamada, K. M., and Kasuga, M., Biosynthesis and glycosylation of the insulin receptor. Evidence for a single polypeptide precursor of the two major subunits. J. biol. Chem. 258 (1983) 10020–10026.

50 Hedo, J. A., and Simpson, I. A., Internalization of insulin receptors in the isolated rat adipose cell. J. biol. Chem. 259 (1984) 11083–11089.

51 Helenius, A., Mellman, I., Wall, D., and Hubbard, A., Endosomes. Trends biochem. Sci. 8 (1983) 245–250.

52 Hernick, C. A., Hamilton, R. L., Spayiani, E., Enders, G. H., and Havel, R. J., Isolation and characterization of multivesicular bodies from rat hepatocytes: an organelle distinct from secretory vesicles of the Golgi apparatus. J. Cell Biol. 100 (1985) 1558–1569.

53 Heuser, J., and Evans, L., Three-dimensional visualization of coated vesicle formation in fibroblasts. J. Cell Biol. 84 (1980) 560–583.

54 Hizuka, N., Gorden, P., Lesniak, M. A., Van Obberghen, E., Carpentier, J.-L., and Orci, L., Polypeptide hormone degradation and receptor regulation are coupled to ligand internalization. A direct biochemical and morphologic demonstration. J. biol. Chem. 256 (1981) 4591–4597.

55 Hubbard, A. L., Wilson, G., Ashwell, G., and Stukenbrok, H., An electron microscope autoradiographic study of the carbohydrate recognition systems in rat liver. I. Distribution of 125I-ligands among the liver cell types. J. Cell Biol. 83 (1979) 47–64.

56 Iacopetta, B. J., Morgan, E. H., and Yeoh, G. C. T., Receptor-mediated endocytosis of transferrin by developing erythroid cells from the fetal rat liver. J. Histochem. Cytochem. 31 (1983) 336–344.

57 Kasuga, M., Carpentier, J.-L., Van Obberghen, E., Orci, L., and Gorden, P., 125I-anti-insulin receptor Fab is internalized by cultured human lymphocytes. Biochem. biophys. Res. Commun. 114 (1983) 230–233.

58 Kasuga, M., Karlsson, F. A., and Kahn, C. R., Insulin stimulates the phosphorylation of the 95,000-dalton subunit of its own receptor. Science *215* (1982) 185–187.

59 Khan, M. N., Posner, B. I., Khan, R. J., and Bergeron, J. J. M., Internalization of insulin into rat liver Golgi elements: evidence for vesicle heterogeneity and the path of intracellular processing. J. biol. Chem. *257* (1982) 5969–5976.

60 King, A. C., and Cuatrecasas, P., Peptide hormone-induced receptor mobility, aggregation, and internalization. New Engl. J. Med. *305* (1981) 77–88.

61 Klausner, R. D., Harford, J., and van Renswoude, J., Rapid internalization of the transferrin receptor in K562 cells is triggered by ligand binding or treatment with a phorbol ester. Proc. natn. Acad. Sci. USA *81* (1984) 3005–3009.

62 Klausner, R. D., van Renswoude, J., Kempf, C., Rao, K., Bateman, J. L., and Robbins, A. R., Failure to release iron from transferrin in a Chinese hamster ovary cell mutant pleiotropically defective in endocytosis. J. Cell Biol. *98* (1984) 1098–1101.

63 Lefkowitz, R. J., Roth, J., Pricer, W., and Pastan, I., ACTH receptors in the adrenal: Specific binding of ACTH ^{125}I and its relation to adrenyl cyclase. Proc. natn. Acad. Sci. USA *65* (1970) 745–752.

64 Marsh, M., Bolzau, E., and Helenius, A., Penetration of Semliki Forest virus from acidic prelysosomal vacuoles. Cell *32* (1983) 931–940.

65 Marshall, S., Green, A., and Olefsky, J. M., Evidence for recycling of insulin receptors in isolated rat adipocytes. J. biol. Chem. *256* (1981) 11464–11470.

66 May, W. S., Jacobs, S., and Cuatrecasas, P., Association of phorbol ester-induced hyper-phosphorylation and reversible regulation of transferrin membrane receptors in HL60 cells. Proc. natn. Acad. Sci. USA *81* (1984) 2016–2020.

67 Mellman, I. S., Steinman, R. M., Unkeless, J. C., and Cohn, Z. A., Selective iodination and polypeptide composition of pinocytic vesicles. J. Cell Biol. *87* (1980) 712–722.

68 Montesano, R., Perrelet, A., Vassalli, P., and Orci, L., Absence of Filipin-sterol complexes from large coated pits on the surface of culture cells. Proc. natn. Acad. Sci. USA *76* (1979) 6391–6395.

69 Nilsson, J., Thyberg, J., Heldin, C.-H., Westermark, B., and Wasteson, A., Surface binding and internalization of platelet-derived growth factor in human fibroblasts. Proc. natn. Acad. Sci. USA *80* (1983) 5592–5596.

70 Orci, L., Carpentier, J.-L., Perrelet, A., Anderson, R. G. W., Goldstein, J. L., and Brown, M. S., Occurrence of low density lipoprotein receptors within large pits on the surface of human fibroblasts as demonstrated by freeze-etching. Expl Cell Res. *113* (1978) 1-13.

71 Orci, L., Perrelet, A., and Gorden, P., Less-understood aspects of the morphology of insulin secretion and binding, in: Recent Progress in Hormone Research, vol. 34, pp. 95–121. Ed. R. O. Greep. Academic Press, New York 1978.

72 Pastan, I., and Willingham, M. C., Receptor-mediated endocytosis: coated pits, receptosomes and the Golgi. Trends biochem. Sci. *8* (1983) 250–254.

73 Pearse, B. M. F., Clathrin: a unique protein associated with intracellular transfer of membrane by coated vesicles. Proc. natn. Acad. Sci. USA *73* (1976) 1255–1259.

74 Peterson, O. W., and van Deurs, B., Serial-section analysis of coated pits and vesicles involved in adsorptive pinocytosis in cultured fibroblasts. J. Cell Biol. *96* (1983) 277–281.

75 Robert, A., Carpentier, J.-L., Van Obberghen, E., Gorden, P., and Orci, L., The endosomal compartment of rat hepatocytes: its characterization in the course of ^{125}I-insulin internalization. Expl Cell Res. *159* (1985) 113–126.

76 Schlessinger, J., Shechter, Y., Willingham, M. C., and Pastan, I., Direct visualization of binding, aggregation, and internalization of insulin and epidermal growth factor on living fibroblastic cells. Proc. natn. Acad. Sci. USA *75* (1978) 2659–2663.

77 Schlessinger, J., Van Obberghen, E., and Kahn, C. R., Insulin and antibodies against insulin receptor cap on the membrane of cultured human lymphocytes. Nature *286* (1980) 729–731.

78 Schlossman, D. M., Schmid, S. L., Braell, W. A., and Rothman, J. E., An enzyme that removes clathrin coats: Purification of an uncoating ATPase. J. Cell Biol. *99* (1984) 723–733.

79 Schmid, S. L., Braell, W. A., Schlossman, D. M., and Rothman, J. E., A role for clathrin light chains in recognition of clathrin cages by 'uncoating ATPase'. Nature *311* (1984) 228–231.

80 Silverstein, S. C., Steinman, R. M., and Cohn, Z. A., Endocytosis. A. Rev. Biochem. *46* (1977) 669–722.

81 Steer, C. J., Bisher, M., Blumenthal, R., and Steven, A. C., Detection of membrane cholesterol by filipin in isolated rat liver coated vesicles is dependent upon removal of the clathrin coat. J. Cell Biol. *99* (1984) 315–319.

82 Steinman, R. M., Mellman, I. S., Muller, W. A., and Cohn, Z. A., Endocytosis and the recycling of plasma membrane. J. Cell Biol. *96* (1983) 1–27.

83   Terris, S., and Steiner, D. F., Binding and degradation of ^{125}I-insulin by rat hepatocytes. J. biol.
       Chem. *250* (1975) 8389–8398.
84   Tycko, B., and Maxfield, F. R., Rapid acidification of endocytic vesicles containing α_2-macro-
       globulin. Cell *28* (1982) 643–651.
85   Ullrich, A., Bell, J. R., Chen, E. Y., Herrera, R., Petruzzelli, L. M., Dull, T. J., Gray, A., Coussens,
       L., Liao, Y.-C., Tsubokawa, M., Mason, A., Seeburg, P. H., Grunfeld, C., Rosen, O. M., and
       Ramanchandran, J., Human insulin receptor and its relationship to the tyrosine kinase family of
       oncogenes. Nature *313* (1985) 756–761.
86   van Deurs, B., Petersen, O. W., and Bundgaard, M., Do coated pinocytic vesicles exist? Trends
       biochem. Sci. *8* (1983) 400–401.
87   Willingham, M. C., Maxfield, F. R., and Pastan, I. H., α_2-macroglobulin binding to the plasma
       membrane of cultured fibroblasts: Diffuse binding followed by clustering in coated regions. J. Cell
       Biol. *82* (1979) 614–625.
88   Willingham, M. C., and Pastan, I., Formation of receptosomes from plasma membrane coated
       pits during endocytosis: Analysis by serial sections with improved membrane labeling and preser-
       vation techniques. Proc. natn. Acad. Sci. USA *80* (1983) 5617–5621.
89   Woods, J. W., Woodward, M. P., and Roth, T. F., Common features of coated vesicles from
       dissimilar tissues: composition and structure. J. Cell Sci. *30* (1978) 87–97.
90   Yamashiro, D. J., Tycko, B., Fluss, S. R., and Maxfield, F. R., Segregation of transferrin to a
       mildly acidic (pH 6.5) para-Golgi compartment in the recycling pathway. Cell *37* (1984) 789–800.
91   Yang-Feng, T. L., Francke, U., and Ullrich, A., Gene for human insulin receptor: localization to
       site on chromosome 19 involved in pre-β-cell leukemia. Science *228* (1985) 728–731.
92   Zick, Y., Rees-Jones, R. W., Taylor, S. I., Gorden, P., and Roth, J., The role of antireceptor
       antibodies in stimulating phosphorylation of the insulin receptor. J. biol. Chem. *259* (1984)
       4396–4400.

The nature and development of steroid hormone receptors*

J. Gorski

Departments of Biochemistry and Animal Sciences, University of Wisconsin, Madison (Wisconsin 53706, USA)

Models of steroid hormone action

Our current conceptualization of the primary steps in steroid hormone (estrogens, androgens, progestins and corticoids) action is shown in figure 1[10]. We now believe that the steroids, which are generally hydrophobic molecules, diffuse through the outer cell membrane and cytoplasm to the nucleus[47]. In the nucleus the steroids bind to their respective receptors which are assumed to be bound with low affinity to some nuclear component[35]. This nuclear component could be DNA, nuclear matrix or some chromatin protein (acceptor) but this has not been proven conclusively. As a result of the steroid-receptor interaction, the receptor's conformation changes dramatically and its affinity for nuclear components becomes much higher[11, 13, 51]. This change in the association of receptor with nuclear components is correlated with, and may lead to the rapid (within minutes) changes in gene expression observed in the target tissues[1, 25, 29]. This general model may apply to all the steroid hormones, vitamin D metabolites and thyroxine, but has not been clearly demonstrated for all such hormones.

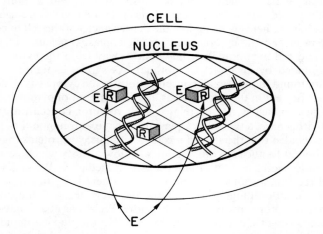

Figure 1. 'New' model of estrogen receptor. R, receptor; E, estrogen; #, nuclear matrix or scaffold; ♪, DNA[10].

The model described above has undergone considerable revision in the last few years. Studies in the 1960s had shown that unoccupied steroid receptors were readily solubilized in cytosolic tissue extracts prepared by conventional homogenization procedures[41]. In contrast, the occupied (steroid bound) receptors were not readily solubilized and were bound in high proportion to nuclear fractions[39]. This led to the proposition of the translocation or two-step model of steroid hormone action[12, 16]. In the translocation model the receptor is initially a cytoplasmic protein. Upon binding steroid the receptor undergoes conformational changes which result in the steroid-receptor complex translocating to the nucleus where it binds to its site or sites of action. Early autoradiography studies were interpreted as supporting the translocation model[16]. While this was the consensus model, not all data were in agreement. Thyroid hormone receptors appeared to be unique as they were only detected in the nuclear compartment of target cells or tissues[30, 36]. The receptor for the vitamin D metabolite, 1, 25-dihydroxy cholecalciferol, was reported to be both cytosolic and nuclear depending on homogenization conditions[45]. Sheridan and associates[38] raised some interesting questions about this model of steroid receptor localization after carrying out autoradiographic studies of estrogen receptor localization. They observed much higher nuclear content of unoccupied estrogen receptor and concluded that receptors exist in an equilibrium between nuclear and cytoplasmic compartments. These autoradiographic studies were complicated, as they required studying uptake of labeled estrogen into intact cells at low temperatures.

In the 1980s immunocytochemistry of the estrogen receptor became possible with the purification of receptor and subsequent preparation of monoclonal antibodies. Initially, such studies supported the translocation model[17, 26], but more recent work indicated all the receptor, with or without ligand, was present in the nucleus[21].

At the same time in our laboratory we were attempting to re-examine this problem using a newer method, cytochalasin-induced enucleation[46]. In this method, illustrated in figure 2, cytochalasin-treated cells are centrifuged through Percoll gradients which allow the denser nuclei of the cell to centrifuge at a faster rate than the cytoplasm. The nucleoplast (nucleus with a rim of cytoplasm and cell membrane) is thus pulled away from the cytoplast (cytoplasm surrounded by an intact membrane). As shown in figure 3, estrogen receptors are found in only small concentrations in the cytoplasts. Unoccupied receptors are present in the nucleoplasts and the amount of cytoplasm remaining in the nucleoplast is not correlated with receptor concentration (fig. 4). Lactic dehydrogenase, a typical cytoplasmic enzyme, is found in the cytoplasts. We have concluded that the steroid receptors are nuclear proteins that without ligand are bound to nuclear components with low affinity which permits their extraction in low ionic strength homogenization media[10, 20]. We believe that all the steroid receptors are likely to have this characteristic. In support of this concept, recent autoradiographic studies have suggested that the unoccupied progesterone receptor is nuclear[9]. Furthermore, Welshons et al.[48] have recently demonstrated, using the enucleation technique, that the glucocorticoid receptor is also nuclear.

An important question that remains to be answered is what keeps the receptor

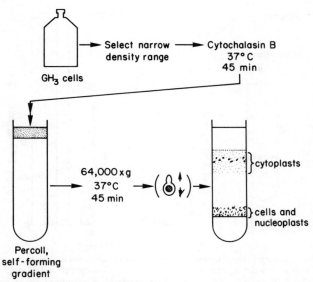

Figure 2. Preparation of cytoplasts from GH$_3$ cells[38a].

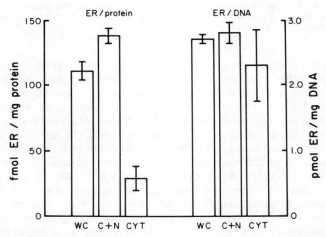

Figure 3. Intracellular distribution of estrogen receptor (ER). GH$_3$ cells were enucleated as described in the text. Estrogen receptor, protein and DNA were measured in untreated density-selected whole cells (WC), in the cell+nucleoplast fraction (C+N) and in the cytoplast fraction (Cyt). In intact cells or cytoplasts, estrogen receptor was measured as specific uptake of ^{3}H-estradiol (151 Ci mmol^{-1}, New England Nuclear) 2 nM in culture medium+DNase (Worthington DPFF 50 µg ml^{-1}) after 30 min at 37°C, with or without 200 nM nonradioactive estradiol. Protein was measured with Coomassie blue staining[25] using the Bio-Rad microassay with bovine α-globulin (Sigma) as the standard. DNA was measured using diphenylamine[26] or the fluorescent dye Hoechst 33258[27]. The figure shows the results of three separate experiments (error bar is the standard error) in which cytoplasts contained 4%, 2.5% and 0.25% contaminating whole cells[46].

70

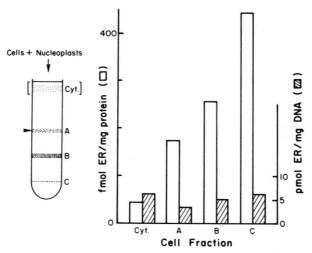

Figure 4. After enucleation, the cytoplasts were saved separately while the cell+nucleoplast layer was further fractionated on a density step gradient, with Percoll (Pharmacia) at 1.04, 1.05, 1.07, 1.08 and 1.10 g ml^{-1}. After 20 min at 400 g, cells were collected at interfaces between 1.05 to 1.07 g ml^{-1} (A), 1.07 to 1.08 g ml^{-1} (B), and 1.08 to 1.10 g ml^{-1} (C). The arrowhead at A indicates where the density-selected whole cells (1.058–1.063 g ml^{-1}) would have been found before enucleation, and the position that the cytoplasts would have occupied is indicated in brackets. Estrogen receptor, DNA and protein were measured in each fraction. Of the cells+nucleoplasts that were recovered after fractionation by density, 15% were recovered at A (which includes the pre-enucleation density), 76% were recovered at B and 9% were recovered at C[46].

in the nucleus. We believe that the receptor is bound to nuclear components. The basis for this is the observation that estrogen receptors immobilized by binding to hydroxylapatite behave like receptor in an intact cell[35]. Estrogen dissociation kinetics, which have been shown to discriminate between transformed and nontransformed receptors, were similar when receptor was immobilized on hydroxylapatite or when freely soluble[35]. On the other hand, the soluble receptor had estrogen binding characteristics that demonstrated positive cooperation as first shown by Notides et al.[28] whereas the immobilized receptor did not[35] (fig. 5). Binding of estrogen in whole cells is not cooperative[49] and thus is similar to the immobilized receptor (fig. 6). In studies of amphibian oocyte nuclei, Feldherr and Pomerantz[8] have concluded that most nuclear proteins are bound to nuclear components although a number of nuclear proteins are readily solubilized. These and other data have led us to propose that the nuclear localization of unoccupied steroid receptor is due to low affinity binding to nuclear components.

Identification of the sites in the nucleus that estrogen receptors bind is difficult. Immunocytochemical studies indicate the receptor is dispersed widely in the nucleus[9]. There is no localization in the nucleolus although less appears in the denser heterochromatin than in less dense enchromatin. The ability to solubilize receptor in low ionic strength buffers suggests a weak association with the nuclear components.

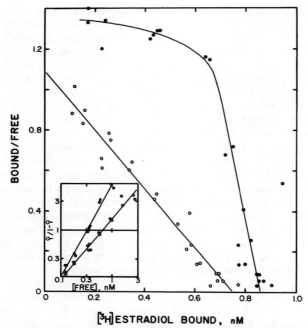

Figure 5. Scatchard plot of equilibrium binding at $0\,°C$ of $[^3H]$-estradiol to native and monomeric estrogen receptors. Inset: Hill plots. (●), Binding to receptor in solution (receptor concentration = 0.85 nM, $S_{0.5}$ = 0.32 nM, n_H = 1.54); (○), Binding to monomeric receptor generated by treatment with 0.4 M KCl and adsorption onto hydroxylapatite (receptor concentration = 0.73 nM, $S_{0.5}$ = 0.64 nM, n_H = 1.01)[35].

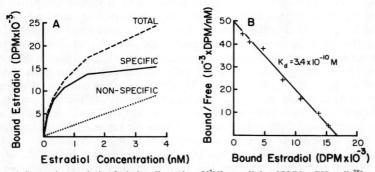

Figure 6. Saturation analysis of whole cell uptake of $[^3H]$ estradiol at $37\,°C$ by GH_3 cells[38a].

Nuclear binding sites

When the unoccupied estrogen receptor binds to an estrogen, a dramatic change occurs in surface properties of the complex indicating a conformational change in the protein[13]. When this interaction occurs at physiological temperatures, further changes are observed including a marked increase in binding to all polyanions, including DNA, RNA, proteins, etc. DNA binding by steroid receptor has been a popular topic for a number of years[51]. Most of such studies have concentrated on the fact that estrogen-receptor complexes bind tightly to DNA but it should be noted they also bind to a number of other substances as well[11]. More recently it has been observed that preferential binding to specific DNA sequences can be detected. Mulvihill et al.[27] were the first to show that a sex steroid receptor (progesterone receptor) bound to a specific sequence of DNA. They reported binding to sequences found upstream of several oviduct genes believed to be regulated by progesterone.

These reports have been followed in greater detail by studies by Yamamoto and associates[31] and others[18] on glucocorticoid receptor binding to mouse mammary tumor virus (MMTV) and metallothionine genes. Specific DNA sequences scattered throughout the viral genome have been shown to have a somewhat higher affinity for the receptor than does the bulk of cellular DNA[31]. A site in the long terminal repeat of the MMTV has been identified as both a receptor binding site and a sequence essential for glucocorticoid activity on transfected MMTV genes[2]. This same sequence appears to have the characteristics of an 'enhancer' sequence[2]. Enhancers are genomic elements which increase the rates of transcription of nearby genes perhaps by increasing RNA polymerase loading[43].

Receptor binding to such sequences is only about 10-fold higher in affinity than binding to other DNA. This is in contrast to the three orders of magnitude higher binding that the lac repressor has for the lac operator compared to nonspecific DNA[24]. It is also in contrast to the recent report from Felsenfeld's group[7] reporting a protein specific for a globin gene expression had $\sim 10,000$ times higher affinity for its specific versus nonspecific sequence. Theoretical discussions of this problem have been presented by Lin and Riggs[24], Travers[42], Ptashne[32] and Von Hippel[44]. The sequence specificity of steroid receptor-DNA binding indicates some sequence homology but surprising numbers of mismatches are permitted[18]. Still, there is a good correspondence between DNA binding and the requirement of the sequence for steroid response. In all such studies purified, naked DNA is being studied. It is well known that in the intact nucleus, DNA is intimately associated with chromatin proteins, both histone and nonhistone proteins. Furthermore, the chromatin may be associated with a fibrillar network called the nuclear matrix or scaffold[15]. While still controversial, the nuclear matrix has been defined as nuclear proteins, insoluble under specific conditions, which may function as sites for replication and transcription[33]. Steroid receptors have been found to remain associated with nuclear matrix after extracting other nuclear components including over 90% of the DNA. Steroid-receptor complexes also bind to nuclear matrix in cell-free experiments[15]. However, if the small amount of residual DNA associated with the nuclear matrix contains operator regions or origins of replication, then DNA binding of the steroid receptor could still be involved in the nuclear matrix.

Chromatin 'acceptor' proteins are also a popular potential site of receptor interaction with chromatin. Ruh and Spelsberg[34] have characterized such proteins from the chick oviduct and others[4] report similar acceptors in rat uteri. A variety of other nuclear components have been identified as receptor binding entities at one time or another[11, 23]. No clear cut evidence is available at this time that clearly defines a specific role for any such nuclear component. This is an active area of research and elucidation of this aspect of steroid hormone action is likely in the near future.

Development of steroid receptors: Ontogeny of estrogen receptors

Regulation of steroid receptor concentration is obviously critical for a cell to be able to respond to a particular steroid hormone. The female sex steroid receptors provide an interesting system for analysis because of their inter-relationships and because they are not needed at all stages of life. Estrogen receptors in two rodent tissues, the uterus and pituitary, are present at very low concentrations at birth[3, 40]. Figures 7 and 8 show the ontogeny of estrogen receptor development in these two tissues. The mouse pituitary has little or no receptor, but measurable levels of receptor are present in the rat uterus[3]. It should be noted that these concentrations are based on whole tissue studies and may reflect either low concentrations throughout the whole cell population or nonrandom distribution limited to certain cell types. An example of the latter case would be the small number of lactotrophs present in the pituitary at birth. The lactotrophs are thought to be a principal target cell for estrogens.

Estrogen receptor concentrations rise until $\sim$ 10–15 days of age in both pituitary and uterus (figs 7 and 8)[3, 40]. After that time, the concentration of receptors remains relatively constant. The table shows that ovariectomy at 1 day after birth has no effect on estrogen receptor ontogeny in rat uteri. Similarly, there is no difference in uterine receptor concentration (fig. 9) in the Ames dwarf mouse whose pituitary is only $\frac{1}{10}$ of normal size and lacks growth hormone, prolactin and thyroid stimulating hormone[37]. It is apparent that the most obvious endocrine regulators do not play a role in regulation of the developmental acquisi-

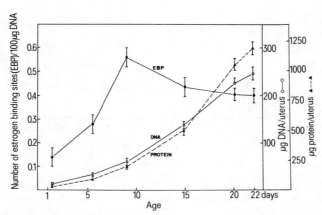

Figure 7. The relationship between the number of estrogen binding sites (EBS) and uterine growth in the immature rat. The number of EBS is expressed as the number of picomoles of estrogen bound. Points on the graph represent the mean ± SEM of 3–4 pooled samples[3].

A comparison of 8- to 10-day-old rat uteri taken from intact and ovariectomized rats. N.S., No significant difference between intact and ovariectomized group. Values represent the mean ± the SEM

Measurement	Intact	Ovariectomized
EBS[a]/uterus (pm)	0.33 ± 0.03	0.35 ± 0.04 N.S.
EBS/100 μg of DNA (pm)	0.53 ± 0.04	0.51 ± 0.05 N.S.
DNA/uterus (μg)	62.0 ± 5.6	60.30 ± 7.5 N.S.
Protein/uterus (μg)	202.0 ± 21.5	195.0 ± 19.6 N.S.
Wet wt (mg)	10.2 ± 0.5	9.9 ± 0.6 N.S.

[a] EBS, estrogen binding sites.

tion of estrogen receptors. A report from Danforth et al.[5] has implicated melatonin in changes of estrogen receptor activity in hamster uteri suggesting this regulator should be examined for its role during receptor development. The importance of receptor development is illustrated by the observation that estrogen receptors in mammary carcinoma are related to the responsiveness of such tumors to endocrine therapy[50]. At the present time we are unaware of any additional data that present any convincing evidence as to how the estrogen receptors are controlled in these tumors.

The progestin receptor is an example of a receptor whose control is largely dependent on another hormone, the estrogens[22]. Progesterone receptor concentration in the uterus is relatively low in rats that have low estrogen levels, such as

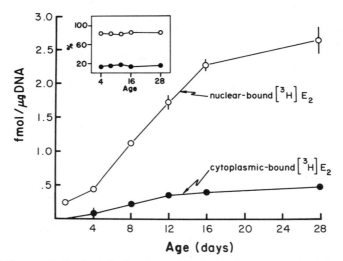

Figure 8. Estrogen binding in pituitaries of postnatal mice. Pituitaries were isolated from mice at the ages indicated. Half of the glands from each age group were incubated with 10 nM [^{3}H]E$_2$ (17β-estradiol), and the other half were incubated with 10 nM [^{3}H]E$_2$ plus 5 μM DES (diethylstilbestrol) for 1 h at 37°C. Cytoplasmic and nuclear bound radioactivities were determined. Bound radioactivity measured in the presence of excess unlabeled competitor was subtracted from the total to estimate specific binding. Each point represents the mean ± SEM of triplicate assays of pooled material. Inset: Receptors measured in the nuclear and soluble fractions were summed, and the ratio found in each fraction was plotted as a function of age[40].

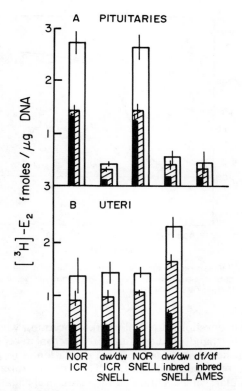

Figure 9. Estradiol receptor levels in pituitaries and uteri of normal and dwarf mice. Nuclear bound and soluble receptor levels were quantitated. Organs from normal and dwarf mice of the corresponding genetic background were assayed in the same experiment. Open bars = total specific binding; hatched bars = nuclear specific binding; closed bars = low-speed supernatant specific binding. A, Pituitary estrogen receptor. Each histogram represents the mean ± SEM of 3 assays (normals and Ames dwarfs) or 6 assays (Snell dwarfs). Pituitaries from normal mice were assayed individually whereas 3 pituitaries from dwarfs were pooled for each hot and hot and cold assay. B, Uterine estrogen receptor. Each histogram represents the mean ± SEM of 3 assays (normals) or 8 assays (Snell dwarfs). In several experiments, uterine receptor values ranging from 1.5 to 4.2 fmoles/g DNA have been measured. These differences are probably due to cyclic variations[37].

immature or ovariectomized females, and is dramatically increased by administration of estrogen[6]. This effect of estrogen is directly on the uterine cells, as can be demonstrated in cell cultures of uterine cells (fig. 10)[19]. It also has been demonstrated in cultures of transformed mammary cells like the MCF-7 cell line[14]. While the details of how estrogen stimulates the production of progesterone receptor await the availability of molecular biological probes, the time course of the induction strongly suggests that a transcriptional site of action is involved. This would be consistent with other estrogen responses.

The state of knowledge concerning the developmental regulation of estrogen versus progesterone receptors are in marked contrast. In the case of the estrogen

76

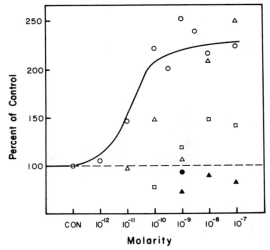

Figure 10. Induction of 130-K protein by various hormones. Cells were cultured for 3 days in media containing various concentrations of 17β-estradiol (○), 16α-estradiol (%), testosterone (□), progesterone (▲) and dexamethasone (●). CON, Control[19].

receptor the major questions are biological, concerning what signals are involved. On the other hand, studies of the progesterone receptor will probably concentrate on the details of transcriptional regulation of its gene and the role of estrogen receptors. Scientific inquiry at all levels of organization, organismic, cellular and molecular, will be required for progress in elucidating the control of steroid receptor development.

*Supported in part by the College of Agricultural and Life Sciences, University of Wisconsin, Madison; NIH Grants HD 08192, CA18110, 5T32HD07259 and National Foundation for Cancer Research.

1   Berridge, M. V., Farmer, S. R., Green, C. D., Henshaw, E. C., and Tata, J. R., Characterisation of polysomes from *Xenopus* liver synthesizing vitellogenin and translation of vitellogenin and albumin mRNAs in vitro. Eur. J. Biochem. *62* (1976) 161–171.
2   Chandler, V. L., Maler, B. A., and Yamamoto, K. R., DNA sequences bound specifically by glucocorticoid receptor in vitro render a heterologous promoter hormone responsive in vivo. Cell *33* (1983) 489–499.
3   Clark, J. H., and Gorski, J., Ontogeny of the estrogen receptor during early uterine development. Science *169* (1970) 76–78.
4   Chuknyiska, R. S., Haji, M., Foote, R. H., and Roth, G. S., Effects of in vivo estradiol administration on availability of rat uterine nuclear acceptor sites measured in vitro. Endocrinology *115* (1984) 836–838.
5   Danforth, D. N., Tamarkin, L., Do, R., and Lippman, M. E., Melatonin-induced increase in cytoplasmic estrogen receptor activity in hamster uteri. Endocrinology *113* (1983) 81–85.
6   Dix, C. J., and Jordon, V. C., Modulation of rat uterine steroid hormone receptors by estrogen and antiestrogen. Endocrinology *107* (1980) 2011–2020.
7   Emerson, B. M., Lewis, C. D., and Felsenfeld, G., Interaction of specific nuclear factors with the nuclease-hypersensitive region of the chicken adult β-globin gene: Nature of the binding domain. Cell *41* (1985) 21–30.

8 Feldherr, C.M., and Pomerantz, J., Mechanism for the selection of nuclear polypeptides in *Xenopus* oocytes. J. Cell Biol. *78* (1978) 168–175.

9 Gasc, J.-M., Renoir, J.-M., Radanyi, C., Joab, I., Tuohimaa, P., and Baulieu, E.-E., Progesterone receptor in the chick oviduct: an immunohistochemical study with antibodies to distinct receptor components. J. Cell Biol. *99* (1984) 1193–1201.

10 Gorski, J., Welshons, W., and Sakai, D., Remodeling the estogen receptor model. Molec. cell. Endocr. *36* (1984) 11–15.

11 Gorski, J., and Gannon, F., Current models of steroid hormone action: A critique. A. Rev. Physiol. *38* (1976) 425–450.

12 Gorski, J., Toft, D., Shyamala, G., Smith, D., and Notides, A., Hormone receptors: Studies on the interaction of estrogen with the uterus. Recent Prog. Horm. Res. *24* (1968) 45–80.

13 Hansen, J., and Gorski, J., Conformational and electrostatic properties of unoccupied and liganded estrogen receptors determined by aqueous two-phase partitioning. Biochemistry *24* (1985) 6078–6085.

14 Horwitz, K.B., and McGuire, W.L., Estrogen control of progesterone receptor in human breast cancer, correlation with nuclear processing of estrogen receptor. J. biol. Chem. *253* (1978) 2223–2228.

15 Barrack, E.R., and Coffey, D.S., Biological properties of the nuclear matrix: Steroid hormone binding. Recent Prog. Horm. Res. *38* (1982) 133–195.

16 Jensen, E.V., Suzuki, T., Kawashima, T., Stumpf, W.E., Jungblut, P.W., and DeSombre, E.R., A two-step mechanism for the interaction of estradiol with rat uterus. Proc. natn. Acad. Sci. USA *59* (1968) 632–638.

17 Jensen, E., Greene, G., Closs, L., DeSombre, E., and Nadji, M., Receptors reconsidered: A 20-year perspective. Recent Prog. Horm. Res. *38* (1982) 1–40.

18 Karin, M., Haslinger, A., Holtgreve, H., Richards, R.I., Krauter, P., Westphal, H.M., and Beato, M., Characterization of DNA sequences through which cadmium and glucocorticoid hormones induce human metallothionein-II$_A$ gene. Nature *308* (1984) 513–519.

19 Kassis, J.A., Sakai, D., Walent, J.H., and Gorski, J., Primary cultures of estrogen-responsive cells from rat uteri: Induction of progesterone receptors and a secreted protein. Endocrinology *114* (1984) 1558–1566.

20 Katzenellenbogen, B.S., and Gorski, J., Estrogen actions on syntheses of macromolecules in target cells, in: Biochemical Actions of Hormones, vol. 3, pp. 187–243. Ed. G. Litwack. Academic Press, New York 1975.

21 King, W.J., and Greene, G.L., Monoclonal antibodies to estrogen receptor localize receptor in the nucleus of target cells. Nature *307* (1984) 745–747.

22 Leavitt, W.W., Chen, T.J., Do, Y.S., Carlton, B.D., and Allen, T.C., Biology of progesterone receptors, in: Receptors and Hormone Action 2, pp. 157–188. Eds B.W. O'Malley and L. Birkbaumer. Academic Press, New York 1978.

23 Liao, S., Liang, T., and Tymoczko, J.L., Ribonucleoprotein binding of steroid-receptor complexes. Nature *241* (1973) 211–213.

24 Lin, S.-Y., and Riggs, A.D., The general affinity of *lac* repressor for E.coli DNA: Implications for gene regulation in procaryotes and eucaryotes. Cell *4* (1975) 107–111.

25 McKnight, G.S., and Palmiter, R.D., Transcriptional regulation of the ovalbumin and conalbumin genes by steroid hormones in chick oviduct. J. biol. Chem. *254* (1979) 9050–9058.

26 Morel, G., Dubois, P., Benassayag, C., Nunez, E., Radaryi, C., Redeuilh, G., Richard-Foy, H., and Baulieu, E.-E., Ultrastructural evidence of oestradiol receptor by immunochemistry. Expl Cell Res. *132* (1981) 249–257.

27 Mulvihill, E.R., LePennec, J.-P., and Chambon, P., Chicken oviduct progesterone receptor: location of specific regions of high-affinity binding in cloned DNA fragments of hormone-responsive genes. Cell *28* (1982) 621–632.

28 Notides, A.C., Lerner, N., and Hamilton, D.E., Positive cooperativity of the estrogen receptor. Proc. natn. Acad. Sci. USA *78* (1981) 4926–4930.

29 O'Malley, B.W., McGuire, W.L., Kohler, P.O., and Korenman, S.G., Studies on the mechanism of steroid hormone regulation of synthesis of specific proteins. Recent Prog. Horm. Res. *25* (1969) 105–160.

30 Oppenheimer, J.H., Schwartz, H.L., Surks, M.I., Koerner, D., and Dillman, W.H., Nuclear receptors and the initiation of thyroid hormone action. Recent Prog. Horm. Res. *32* (1976) 529–565.

31 Payvar, F., DeFranco, D.G., Firestone, L., Edgar, B., Wrange, D., Okret, S., Gustafsson, J.-A., and Yamamoto, K.R., Sequence-specific binding of glucocorticoid receptor to MMTV DNA at sites within and upstream of the transcribed region. Cell *35* (1983) 381–392.

78

32 Ptashne, M., DNA-binding proteins. Nature *303* (1983) 753–754.
33 Robinson, S. I., Small, D., Idzerda, R., McKnight, G. S., and Vogelstein, B., The association of transcriptionally active genes with the nuclear matrix of the chicken oviduct. Nucl. Acids Res. *11* (1983) 5113–5130.
34 Ruh, T. S., and Spelsberg, T. C., Acceptor sites for the oestrogen receptor in hen oviduct chromatin. Biochem. J. *210* (1983) 905–912.
35 Sakai, D., and Gorski, J., Estrogen receptor transformation to a high-affinity state without subunit-subunit interactions. Biochemistry *23* (1984) 3541–3547.
36 Samuels, H., and Tsai, J., Thyroid hormone action in cell culture: Demonstration of nuclear receptors in intact cells and isolated nuclei. Proc. natn. Acad. Sci. USA *70* (1973) 3488–3492.
37 Sartor, P., Slabaugh, M., Sakai, D., and Gorski, J., Estrogen receptor development in the absence of growth hormone and prolactin: Studies in dwarf mice. Molec. cell. Endocr. *29* (1983) 91–99.
38 Sheridan, P. J., Buchanan, J. M., and Anselmo, V. C., Equilibrium: the intracellular distribution of steroid receptors. Nature *282* (1979) 579–582.
38a Shull, J. D., Welshons, W. V., Lieberman, M. E., and Gorski, J., The rat pituitary estrogen receptor: role of the nuclear receptor in the regulation of transcription of the prolactin gene and the nuclear localization of the unoccupied receptor, in: Molecular Mechanism of Steroid Hormone Action. Ed. V. K. Moudgil. Walter de Gruyter and Co., New York 1985.
39 Shyamala, G., and Gorski, J., Estrogen receptors in the rat uterus: Studies on the interaction of cytosol and nuclear binding sites. J. biol. Chem. *244* (1969) 1097–1103.
40 Slabaugh, M. B., Lieberman, M. E., Rutledge, J. J., and Gorski, J., Ontogeny of growth hormone and prolactin gene expression in mice. Endocrinology *110* (1982) 1489–1497.
41 Toft, D., Shyamala, G., and Gorski, J., A receptor molecule for estrogens: Studies using a cell-free system. Proc. natn. Acad. Sci. *57* (1967) 1740–1743.
42 Travers, A., Protein contacts for promoter location in eukaryotes. Nature *303* (1983) 755.
43 Treisman, R., and Maniatis, T., Simian virus 40 enhancer increases number of RNA polymerase II molecules on linked DNA. Nature *315* (1985) 72–75.
44 Von Hippel, P. H., Bear, D. G., Morgan, W. D., and McSwiggen, J. A., Protein-nucleic acid interactions in transcriptions: A molecular analysis. A. Rev. Biochem. *53* (1984) 389–446.
45 Walters, M., Hunziker, W., and Norman, A., Unoccupied 1, 25-dihydroxyvitamin D_3 receptors. Nuclear/cytosol ratio depends on ionic strength. J. biol. Chem. *255* (1980) 6799–6805.
46 Welshons, W. V., Lieberman, M. E., and Gorski, J., Nuclear localization of unoccupied oestrogen receptors. Nature *307* (1984) 747–749.
47 Welshons, W. V., and Gorski, J., Nuclear location of estrogen receptors, in: The Receptors, Vol. 5. Ed. P. M. Conn. Academic Press, New York, in press (1986).
48 Welshons, W. V., Krummel, B. M., and Gorski, J., Nuclear localization of unoccupied receptors for glucocorticoids, estrogens and progesterone in GH_3 cells. Endocrinology *117* (1986) 2140–2147.
49 Williams, D., and Gorski, J., Equilibrium binding of estradiol by uterine cell suspensions and whole uteri in vitro. Biochemistry *13* (1974) 5537–5542.
50 Wittliff, J. L., Steroid-hormone receptors in breast cancer. Cancer *53* (1984) 630–643.
51 Yamamoto, K. R., and Alberts, B. M., Steriod receptors: Elements for modulation of eukaryotic transcription. A. Rev. Biochem. *45* (1976) 721–746.

Receptor ontogeny and hormonal imprinting

G. Csaba

Department of Biology, Semmelweis University of Medicine, POB 370, H–1445 Budapest (Hungary)

Continuous interaction with the environment is a basic feature of cell life. Precise recognition of the environment is a prerequisite for cell function and survival. The environment of the unicellular organism is the external world. Transition from unicellular to multicellular forms of life involves dramatic alterations in the recognition system since, while the environment of the multicellular organism is still the external world, that of the single cell may be, depending on its localization, either the external or the internal milieu[15, 16]. The at least partial 'internalization' of the environment at the level of multicellularity results in specialization of the recognition system (for the needs of the community). While the developing nervous system acquires the ability to receive scores of different signals and to control the entire (multicellular) organism, the primordial chemical system of signal recognition acquires different functions which become integrated into the complexity of the functions of the organism.

Differentiation of 'self' from 'non-self' (foreign) seems to exist at all levels of phylogenesis, because even unicellular organisms combine and form colonies exclusively with their own kind, and either escape or devour the organisms recognized as 'foreign'. From this basic mechanism develops at the multicellular level the marker–receptor recognition system, whose functions are to furnish morphogenesis and to maintain homeostasis of the immune system. Thus the immune system of multicellular organisms screens both the external and the internal environment for chemical (molecular) signals emitted by 'self' and 'foreign' cells. The second complex of the chemical recognition systems is represented by the pheromone system, which controls the relations between individuals, and is oriented exclusively towards the external environment. Last but not least, the endocrine system of multicellular organisms evolves from the primitive chemical recognition system of the unicellular organism. The endocrine system coordinates cellular functions initially in the absence of, but later in collaboration with, the nervous system. Since the endocrine system operates inside the multicellular organism, the recognitive function of the cells controlled by it is oriented towards the internal environment.

Analysis of the recognition system of multicellular organisms from the point of view of environmental relations has revealed that essentially four sub-systems operate in that system, of which one interacts exclusively with the external environment (pheromone system), two can process both external and internal signals (nervous system and immune system), and one responds exclusively to internal signals (recognition part of endocrine system). The operation of the sub-systems is genetically encoded, within the limits set by the needs and potentialities of the species. However, it appears that only the sub-system which responds exclusively to external signals (pheromone system) is encoded in every detail. In each condition in which recognition of the internal environment, i.e. recognition of the individually varying 'self' is decisive, normal regulation presupposes the operation of the controlling and controlled factors in a complementary code system, and adjustment (adaptation) of the two factors to one another. In all probability, since individual variation tends to increase with the progression of evolution, adaptation, too, obtains an increasing importance. This is reflected by the immune differentiation of 'self' from 'foreign'. The immune system recognizes as 'self' all structures which were present in the course of its development, and as foreign all those with which it has interacted in its fully developed form, regardless of whether the latter structures are useful or noxious for the organism. The same applies to behavioral imprinting, which plays a decisive role in the fundamental adaptation of affective responses. In this light it appears logical that the endocrine system, too, has to adapt itself for self (individual milieu), because its entire function is oriented on the internal environment.

The endocrine system can be subdivided into two main functional entities, the hormones and the hormone receptors. Supposing that these structures are not stable from the very beginning, but adaptable to one another, the conclusion lies close at hand that either is capable of adapting to the other[12]. From the phylogenetic point of view the hormones appear to be more dynamic structures than the receptors[63], for they may vary considerably by mutation. For example, only about 38% of the amino acid components of the vertebrate hormone insulin have persisted in an unchanged form throughout the course of about 500 million years' evolution, whereas the receptors of the hagfish, which took a different evolutionary path about 500 million years ago, are still able to bind vertebrate insulin[76,77]. On the other hand, ontogenetic investigations[12] suggest that the hormones have been stable structures ever since they appeared, and adaptability seems to be primarily the property of the receptors.

Receptor alterations during ontogenetic development

The hormone-binding capacity of receptor-bearing cells changes greatly in the perinatal period. The glucagon sensitivity of embryonic rat liver cells is only 1% of the adult value after 15 days of prenatal life, and 23% of it after 21 days. Prenatal hepatocellular insulin binding capacity is 11% and 45% relative to the adult value at 15 and 21 days of prenatal life, respectively[9]. The luteinizing hormone (LH) receptors of the rat testicle can be detected as early as at 15.5 days of prenatal life, show a considerable increase in binding capacity at 18.5 days, and become fully active by the 5th postnatal day[92]. The receptors of the follicle stimulating hormone (FSH) appear in the rat embryo at 17.5 days of intrauterine life. Their binding capacity remains low until 19.5 days, but tends to increase markedly from 20.5 days on. The hepatic insulin-binding capacity of the human fetus tends to increase gradually, whereas the number of somatomedine receptors shows practically no change[85]. At the same time, the number of insulin receptors has been found to decrease in adulthood, from the neonatal value of 44,600/cell to 7100/cell[90], and this quantitative decrease has also been substantiated by other experimental observations[4]. The somatotropic receptors of the female rat increase 5 to 6 times in number from the 8th to the 28th postnatal day[73]. The hepatic glucagon receptors of the guinea pig fetus are 3 times more numerous at 58 than at 65 days of age, and relatively less numerous in adulthood[62]. Prenatally the glucagon receptors do not differ in amount between 58 and 65 days of embryonic life, but their binding affinity is relatively greater at 65 days, although their number is about 50% less than in adulthood. The beta adrenergic receptors of chickens are less numerous at 9 days of age than between 4, 5 and 7.5 days, and still fewer at 12–16 days[1]. While adrenalin is able to act on neonatal rat hepatocytes in culture, insulin has hardly any influence on them[17]. The triiodothyronine (T_3) binding capacity of the rat is considerably greater in early prenatal life than at the newborn age[42], and is lower in adulthood than neonatally[41].

The above experimental observations suggest that the number of receptors, and to a lesser degree their affinity for the hormone, does change during the perinatal period. It cannot be stated that the number, or binding capacity, of the receptors tends to increase from neonatal to adult age, nor is there firm evidence to the contrary. Nor can it be stated that an increase or decrease in binding capacity with increasing age would apply to any given hormone or receptor. It appears that receptor numbers vary with the degree of maturity at birth, probably also with the sex, and almost certainly with the type of the hormone, too, because certain hormones which play a morphogenetic role, prenatally, are known to switch over to another function postnatally.

In view of this it seems logical that the receptors of morphogenetic hormones already appear in fetal life, and are probably more numerous and display a greater binding affinity in that period than after birth[64]. However, the number of receptors is in itself not conclusive evidence of the intensity of hormone action for, although the receptor structure may itself be present prenatally, its postreception mechanism may be out of operation to protect the developing cell against an undesirably strong hormonal influence[61]. Perinatal changes in the number and affinity of the receptors seem to be associated with the gradual stabilization of the originally unstable (plastic) binding site (e.g. hormone receptor stabilization takes about one month in the rat). In principle, stabilization could be the result of a spontaneous – genetically encoded – differentiation associated with that of the receptor-bearing cell membrane or cytosol, but it may also be an adaptation phenomenon which is facilitated by the initial perinatal plasticity of the receptor structure. Perinatal adaptation is in all probability required for the establishment of the final receptor number and structure.

Why is receptor adaptation necessary? An approach based on information theory

The main components of an information system are the signal emitter, the signal receiver, and the channel through which the signal (information) passes from emitter to receiver. Information transfer presupposes that either the emitter and receiver operate in the same code system, or the receiver is dynamic enough to receive and process practically all signals, regardless of their code system[12]. The receiver structures of unicellular organism are probably of the latter type, inasmuch as their dynamic membrane seems to be able, under the influence of the signal molecule, to present a complementary interacting pattern to a great variety of signal molecules. Receptor formation by such a mechanism is, however, definitely not characteristic of multicellular (endocrine-system-possessing) organisms, in which exchange of information with a closed internal environment presupposes the encoding of the signal emitter and signal receiver by the same genetic coding system. Thus it appears that at the level of the organisms possessing an endocrine system, the signal emitter and signal receiver necessarily operate within the same (known) complementary code system.

Nevertheless, taking into consideration that although the signal emitter (endocrine cell) and signal receiver (target cell) carry the same genetic information, different details of them may operate in the different cell types; it is by no means certain that the interrelationships of the gene details responsible for coding the emitter and receiver are heritable. In other words, although the internal environment is heritable, its hereditary transmission may not a priori be coordinated with the heritage of the cells it controls. There is reason to postulate that the endocrine changes associated with certain pathological or nearly pathological phenomena represent only a marginal sector of the variations which fall into the normal range, and account for the great variability of the receptor–hormone relationship. In this light the signal emitter and signal receiver, and the code of their operation are deterministic, but adequate operation of information transfer requires tuning of the emitter and receiver to one another, to ensure the functioning of a living system. Perinatal adaptation, which ultimately results in the establishment of the final, reproducible, receptor number and, probably, receptor affinity, seems to be the underlying mechanism of mutual adjustment[11, 13].

The third main component of the information system is, apart from the code, the channel, which in higher organisms is represented by the circulation or in certain cases, e.g. in the case of paracrine secretion, by the intracellular space. Through this channel pass various signal molecules, along with molecules with other (non-signal) functions, which may be capable or incapable of binding to receptor structures. Thus the target cell (i.e. its receptor) is obliged to select from among a mass of diverse molecules those carrying the appropriate information, and to differentiate them from the general 'noise'. Selection is difficult, because the hormones form hormone families[6, 7] (partly because it is easier to 'use' a modified version of an established signal molecule, partly because the hormone synthesizing cells are in many cases similar) and the more similar the signals, the more difficult it is to distinguish one from the other.

Exactly for this reason it is imperative that the target cell adapts to the specific hormone while the receptor is still plastic, in order that certain factors which avert the confusion of similar signals can take effect. One such factor is probably the circumstance that the specific hormone and the related molecules do not simultaneously appear in the perinatal period at the concentrations required for receptor adaptation[54, 56, 57, 71, 78]. Furthermore, the binding (transport) proteins of the circulating blood may also play a role in the control of the hormone concentration. On the other hand, as the availability of the receptor for the hormone is limited, the delay of receptor maturation relative to hormone availability also facilitates precise adaptation. At all events, this concept of mutual adaptation of receptor and hormone presupposes a precise programming of the critical events (timing of specific hormone peak and receptor maturation).

The chemical code system of the living organisms is, despite the existing similarities between hormones, fairly variable. In the case of polypeptide hormones, the amino acid sequences appear to be fairly simple, but their association into secondary, tertiary and quaternary structures increases the intricacy of the molecular structure. Since the closest relationship (in terms of amino acid sequence) has been demonstrated exactly between polypeptide hormones, these sequences seem to account for the specificity of their code. The amino acid-type, steroid-type and fatty acid-type hormones arise by different codes, and their specificity is associated with certain determinant groups. The lower the phylogenetic level of the organism, the less specific appear to be its receptors, and the greater are the overlaps of signal molecules (e.g. of polypeptide hormones) on one another's binding sites[22, 59]. Probably the phylogenetic process of receptor formation is being reproduced during the ontogenetic development of the receptor, and adaptation is required to impress the 'image' of the specific hormone on the 'memory' of the cellular receiver structure.

Experimental evidence for hormonal imprinting

An experimental approach to the substantiation of neonatal adaptation (imprinting) calls for the creation of either hormone deficiency or hormone excess during the critical period. The effect of both treatments is assessed in adulthood.

Polypeptide hormones

The thyrotropic hormone (TSH) is a polypeptide hormone. Treatment of newborn rats with thyroxine (T_4) for several days depresses hypophyseal TSH-production, thus giving rise to so-called neonatal hyperthyroidism[55]. Animals so treated while newborn show no decrease in the basic thyroid hormone level in adulthood, but no increase either in T_3- or T_4-production in response to TSH[34]. It appears that normal receptor development requires the presence of TSH, because absence of TSH in the critical neonatal period accounts for depression of adult response to TSH. On the other hand, neonatal exposure to a massive dose of TSH also results in depression of adult thyroidic response, not only in rats[33], but also in chickens[49]. The period required for the normal course of imprinting can be fairly well assessed in chickens; these are more mature than the rats at birth, and, accordingly, the development of their hypothalamic-pituitary-thyroid axis also is more advanced at hatching[75]. While the chick embryo is irresponsive to TSH at 8 days of prenatal life, it shows a lifelong alteration in receptor responsiveness when exposed to TSH at 12 days[79], after the above axis has developed. For the rat, the optimal hormone dose is 20–50 IU; doses above or below that range are depressive or indifferent rather than stimulatory[80].

Testing follicle stimulating hormone (FSH) on the gonads of day-old chicks has shown that neonatal exposure to FSH (of chicks) amplified the FSH-receptor[21].

84

Certain organs of the chickens so treated developed more intensively also without reexposure, owing in all probability to an increased responsiveness of the target cells to the endogenous hormone. Gonadal response proved to be still greater after reexposure to FSH in adulthood[84].

A single neonatal treatment with vasopressin unequivocally accounted for an increase in adult response to vasopressin[38].

Mice exposed to met-encephalin neonatally on a single occasion showed a considerable increase in sensitivity to opioids in adulthood[84].

The effect of a single neonatal exposure to insulin can be assessed from alterations in the hepatic hormone binding and hepatocellular responsiveness of the adult[66]. There is, however, a sex variation, for neonatal insulin exposure enhances adult insulin binding in females, but depresses it in males, and has an adequate effect on the blood glucose level, too (fig. 1).

Imprinting by related hormones; the disturbing effect of 'noise'

TSH and the gonadotropins are related, in that they have a common alpha subunit and their beta subunits differ only slightly, just to a degree which furnishes specificity of action[91]. TSH-gonadotropin binding overlaps can occur in adulthood without any appreciable consequence in function[2,3,83], but they have far-reaching consequences in the critical perinatal period.

Assessment of the effects of exposure to FSH or TSH on different parameters of chicken gonads has shown that the two hormones develop qualitatively the same action when applied to day-old chickens[20]. Although FSH has a greater influence on the germinal elements than TSH and, vice versa, the latter has a more powerful action than the former on the interstitium, each has an appreciable effect on the target cells of the other. Quantitative measurements have

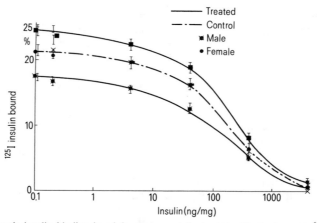

Figure 1. Hepatic insulin binding in adult rats neonatally treated with the hormone[66]. While the binding values of the control rats practically did not differ between the sexes, the experimental males showed a decrease, the females an increase, in hepatic insulin binding capacity.

revealed that the gonadal action of TSH was also superior to that of FSH at lower dose levels. This cannot be ascribed to a contamination, on the one hand because the preparations used were very pure, on the other because the effect of contamination cannot supersede hormone action. The functional overlaps could be substantiated by light and electron microscopic studies which showed that the two hormones stimulated the function of the same organelles[40,86,87]. Evidence of TSH-FSH functional overlap at the neonatal age has strongly suggested that these hormones bind to the same receptors in that period. This has prompted us to investigate the influence of structurally similar (related), but functionally different hormones on receptor development.

Rats[33] and chickens[49] neonatally exposed to a single dose of gonadotropin had a decreased T_4-level in adulthood. Thyroidic response to TSH was weak, to judge from a negligible rise in the T_4-level (fig. 2).

Treatment of newly-hatched chicks with TSH amplified the gonadotropin receptors to a similar degree to FSH itself, to judge from morphological and functional (hormone level) evidence of a greater response to FSH on reexposure[49]. TSH proved to be in several respects more active than FSH, whereas it failed to amplify the receptors for itself, to judge from the experimental observation that response to TSH of adult chickens neonatally preexposed to TSH did not differ from that of the not-preexposed control. The gonadal response of

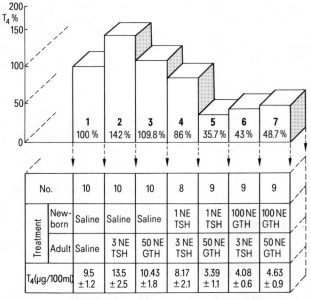

Figure 2. T_4-production of rats treated neonatally on a single occasion with gonadotropin (GTH) or thyrotropin (TSH), with and without reexposure to GTH- or TSH in adulthood[33]. While the control rats responded well to TSH, of the neonatally preexposed rats only those treated with TSH showed a measurable, but considerably reduced response to challenge in adulthood, and those preexposed to GTH were irresponsive to both TSH- and GTH-reexposure.

neonatally TSH-treated rats was qualitatively similar to that of the chickens, but quantitatively less conspicuous.

Neonatal influence of exogenous TSH or FSH on the receptors also alters adult sexual behavior[24], although the functional overlap of TSH is of a lesser degree than in the case of morphological or hormonal indexes.

The polypeptide hormones vasopressin and oxytoxin are also related molecules, differing from each other only in two amino acid components. No functional overlap between them is demonstrable in adulthood, but neonatal exposure of rats to oxytocin amplifies the receptors for vasopressin as well as for oxytocin itself and this effect is still demonstrable in adulthood[38].

Overlapping imprinting by steroid hormones

Neonatal treatment with sexual steroids influences the hypothalamic receptors and through them the adjustment of sexual constitution[5, 8, 10, 50, 53]. However, target cells other than hypothalamic may respond to steroids at receptor level, and imprinting by overlapping steroid hormones can also occur.

Rats treated neonatally with a single dose of diethylstilbestrol (DES) showed at 6 weeks of age a decrease of about 66% in uterine estradiol receptor numbers without, however, any change in receptor affinity[29]. A single neonatal exposure to allylestrenol instead of DES accounted for a decrease of about 50% in the number of estradiol binding sites, also without causing any alteration in their affinity (fig. 3). Neonatal exposure to DES or allylestrenol also gave rise to a considerable alteration (reduction) of adult uterine DES-binding capacity.

Primary exposure to DES in early pregnancy (at days 9 and 14 of gestation) instead of the newborn age resulted in death or stillbirth of all fetuses. The

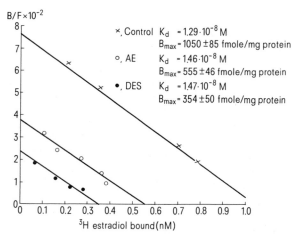

Figure 3. Uterine estradiol binding capacity of female rats treated neonatally with diethylstilbestrol (DES) or allylestrenol (AE). Allylestrenol accounted for an about 50% decrease in estradiol binding, and DES for loss of about two thirds of the estradiol receptors[29], as assessed by Scatchard plots.

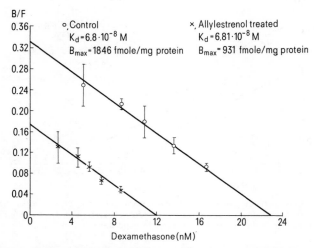

Figure 4. Decrease in thymic glucocorticoid (dexamethasone) receptor numbers in adult rats treated with allylestrenol on a single occasion in the neonatal age[68].

offspring of the allylestrenol-treated females were born alive, but no change in the uterine binding capacity was demonstrable relative to the control in that case. At the same time, a decrease of about 50% was observed in the number of the thymic glucocorticoid receptors[69].

Neonatal imprinting with diethylstilbestrol or allylestrenol may also influence animal behavior. Both compounds depress considerably the adult sexual activity of males, but only DES, the stronger acting of the two, has a similar effect on females[25].

Rats neonatally treated with glucocorticoid (dexamethasone) showed a decrease of about 33% in thymic glucocorticoid receptor numbers in adulthood[68] (fig. 4).

The steroid structure is very widely distributed in the living world; it is demonstrable in both plant and animal organisms. Thus the steroid structure is characteristic not only of steroid hormones, but also of a variety of materials which occur in the natural environment of man, and act on him either advantageously as drugs, or noxiously as carcinogens. The basic structure of the digitalis derivatives and of benzpyrene is also steroid-like.

Rats treated neonatally on a single occasion with digoxin or ouabain[27] responded to reexposure in adulthood with an increase in the blood digoxin level (after digoxin treatment), i.e. by decreased elimination of digoxin. A single neonatal treatment with benzpyrene[26] accounted for a 33% decrease in adult thymic glucocorticoid reception (fig. 5). Neonatal exposure to ouabain had no influence on adult thymic glucocorticoid binding. Conversely, however, neonatal triamcinolone treatment[67] depressed adult myocardial ouabain binding and ouabain sensitive ATP-ase activity by about 33%. Neonatal allylestrenol treatment did not influence the ouabain binding of myocardial cells, while

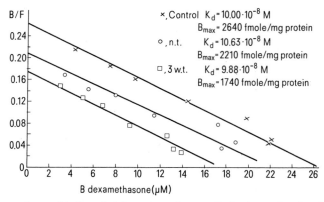

Figure 5. Dexamethasone binding of adult rats treated neonatally (n. t.) or at 3 weeks of age (3. w. t.) with a single dose of benzpyrene[26]. The number of glucocorticoid receptors was considerably reduced by the neonatal exposure, and to a still greater degree by exposure at 3 weeks of age.

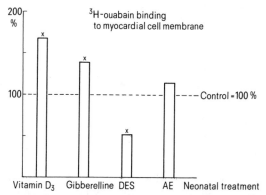

Figure 6. While neonatal allylestrenol treatment had no influence on myocardial ouabain binding in adulthood, neonatal imprinting with DES decreased it considerably, and imprinting with gibberelline or vitamin D_3 increased it to a similar degree[70]. x, Significance to control; $p < 0.01$.

neonatal DES imprinting decreased, gibberelline or vitamin D_3 treatment increased it to a similar degree[70] (fig. 6).

Imprinting with amino acid type hormones

A single neonatal treatment with catecholamines (epinephrine, isoproterenol, dopamine) alters considerably the adrenergic vascular response of the adult rat. Isoproterenol alters adult response to norepinephrine and to vasopressin as well[39].

A single treatment with melatonin at the newborn age accounted for about a

25% increase in the adult T_4-level without any further exposure to the hormone. Since melatonin is a thyroid inhibitor, neonatal treatment with it presumably gives rise to desensitization of the thyroidic melatonin receptors[35]. At the same time melatonin inhibits the action of TSH on the thyroid gland.

The importance and sensitivity of imprinting

The foregoing considerations permit the following conclusions:

1. Presence of the hormone is an essential prerequisite of the amplification (imprinting) of the receptor.

2. Related hormone molecules, whose chemical structure differs from that of the specific hormone but not to such an extent which would prevent their binding to its receptor, induce either an adequate imprinting characteristic of the hormone proper, or a different (faulty) imprinting.

3. By interfering with the normal course of imprinting, the related molecule biases response to the specific hormone without, however, adapting the receptor for itself; thus no nonspecific response occurs in adulthood. The exception is the case in which the cell membrane is artificially damaged[36] during imprinting with the related hormone; in such case the related molecule involved in imprinting acquires a lasting advantage over the adequate hormone (fig. 7).

4. In principle, adaptation of the hormone and receptor to one another is optimal if it takes place without 'foreign' interference, although faulty imprinting could presumably be corrected by 'artificial' interference.

5. The specific hormone itself can damage the receptor, if it appears at an inappropriate time and concentration (fig. 8).

Thus hormonal imprinting is a species-dependent, time-dependent (and probably also sex-dependent) biological phenomenon, whose effects may be different exactly for the reasons outlined above. Its effect is portrayed above all by the numbers of (maximal) evokable receptors, for no appreciable affinity changes have been demonstrable by receptor kinetic analysis. *Imprinting is in all prob-*

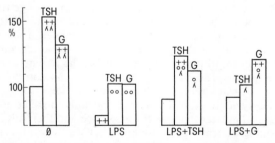

Figure 7. T_4-production in response to thyrotropin (TSH) or gonadotropin (G) exposure of adult rats treated neonatally on a single occasion with endotoxin (LPS) or LPS + hormone[36]. The increase over the control was significant at $p < 0.05 = +$; $p < 0.01 = + +$ level, over the rats treated exclusively in newborn age at $p < 0.01 = + +$; $p < 0.05 = 00$ level, and the significance of inter-group difference between the TSH- and G-exposed rats was $p < 0.05 = ʎ$, $p < 0.01 = ʎʎ$. The rats neonatally exposed to LPS + G were more responsive to G- than to TSH-challenge.

90

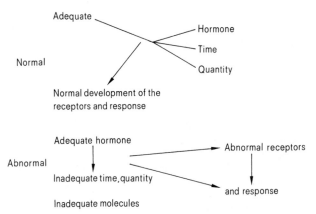

Figure 8. Hormonal imprinting is an indispensable prerequisite of receptor maturation. If imprinting is accomplished by a 'foreign' (non-adequate) hormone, or at inadequate time or concentration, a qualitatively or quantitatively faulty receptor 'behavior' becomes established at a later stage of development.

ability not the exclusive property of hormone receptors, as non-hormone molecules acting at receptor level can also induce it (in their own receptors or in related binding structures). This has been demonstrated in the case of glucose-like molecules, such as glucosamine and mannose[19], and of drugs, chemicals, etc.

Once induced and established, imprinting persists for the lifetime of the organism, as has been demonstrated in the FSH-TSH, and vasopressin-oxytocin systems. In the unicellular *Tetrahymena,* imprinting has been shown to persist by transmission from one cell generation to the other[37]. This is, in all probability, the case also in higher organisms, for in these, too, cells treated in the neonatal age represent distant ancestors of those which are present in adulthood. When treatment is performed at the newborn age (when imprinting occurs), and its effect is measured in adulthood, many new generations arise in the time interval between, which obviously 'remember' the event of imprinting. The 'heredity' of imprinting seems to persist still longer. The progeny generation of rats treated with insulin neonatally, and mated in the adult age, showed the same pattern of imprinting after maturity as the parent generation, in that insulin binding was decreased in the males, and increased in the females[28]. Since the F_1 rats had not been treated with insulin neonatally, their changed insulin binding was clearly acquired by hereditary transmission. The effect of parental imprinting was still greater in those F_1-rats which had themselves been exposed to insulin at the newborn age. A similar effect was also observed if only one parent of the two had been treated, but the binding capacity (affinity) of the receptor was weaker in such cases. Parental imprinting was no longer demonstrable in the F_2 generation.

Is Haeckel's principle valid in ontogenetic development of hormone receptors?

The receptors of the adult rat are highly selective, and can therefore (functionally) differentiate the closely related hormones FSH and TSH from one another. This selectivity is not yet present at the newborn age, when the rat gonads respond equally to FSH and TSH. At a lower phylogenetic level, in the frog, FSH-TSH overlap can also occur in adulthood[59]. Treatment of adult male frogs with chorionic gonadotropin characteristically elicits ejaculation (this was formerly utilized for pregnancy testing). TSH has the same effect on the adult male frog, although at a higher dose than FSH[22]. While this effect of TSH is inferior to that of chorionic gonadotropin, it is certainly superior to that of the luteinizing hormone (fig. 9). Moreover, TSH potentiates the action of the gonadotropic hormone in the frog[23]. This does not, of course, mean that hormonal imprinting can also be induced at the adult age at the lower levels of phylogenesis; it signifies rather that the receptors of phylogenetically lower organisms are as incapable of differentiating related hormones from one another in adulthood as those of higher organisms are in the perinatal period, but the effects of hormone overlap are dissimilar at the two phylogenetic levels.

Hormonal imprinting in cell lines

Cell lines have generally been developed from mammalian cells which had been subject to hormonal imprinting in vivo. Nevertheless, long-term maintenance by serial passages in all probability results in loss of the 'memory' of the event, because imprinting usually takes place in mammalian cell lines exposed to a hormone in vitro, to judge from an increase in hormone binding after a few days[14]. The effects and the overlap of TSH and FSH can be reliably studied in

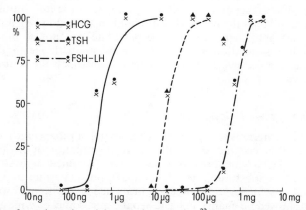

Figure 9. Effect of gonadotropin and thyrotropin on the frog[22]. To provoke ejaculation, the required FSH-LH dose was one order higher than the active those of TSH, and HCG was superior in action to both. It follows that at the phylogenetic level of the frog, the ralated hormone TSH may overlap the specific hormone also in adulthood, and may even develop a stronger action, as in the case of FSH-LH.

Chinese hamster ovary (CHO) cell cultures. TSH enhances CHO cell division to the same extent as FSH[45]; pretreatment with FSH amplifies the CHO-cell receptor not only for itself, but also for TSH; and vice versa, TSH amplifies it for both itself and FSH[47]. By analogy with perinatal imprinting, TSH is superior in action to FSH, and amplifies the receptor to a greater degree for FSH than for itself, also in vitro.

Likewise, insulin, too, can induce imprinting in cell culture[46], at a very low concentration (10^{-13} M) and in a relatively short time (1 h). However, the insulin imprinting induced in cultured cells is not equivalent to that induced in the living (animal) organism, because it expires, or becomes disturbed, after a certain time. This can probably be explained by the circumstance that cultured mammalian cells have necessarily been subject to hormonal imprinting in vivo, and their primary exposure in culture brings about a revival rather than a primary induction of imprinting. It should also be taken into consideration that while the effect of hormonal imprinting is maintained by the endogenous hormone in the living organism, this is not the case in culture.

Cell-cell transmission of hormonal imprinting in cell lines

Fletcher and Greenan[58] reported in 1985 that although only 25–30% of ovarian granulosa cells was able to bind hCG in culture, release of protein kinase did take place in a greater percentage of the cell population. This suggested transmission of the hormone-mediated information to those cells, which had not themselves bound the hormone. This implication prompted investigations into the cell-cell transmission of hormonal imprinting. Evidence of transmission of insulin, TSH, FSH and LH imprinting to cells not interacting with these hormones was obtained in cultures of the Chang liver and Chinese hamster ovary cell lines[31], and in cultures of the unicellular *Tetrahymena*[32] as well. However, while in the case of *Tetrahymena* the culture medium played a key role in transmission, indicating an extracellular secretion of the imprinting factor, in the case of the cell lines the underlying mechanism seemed to be a direct cell-cell contact. In this light there is reason to postulate that not so much the gap junctions as the interrelationships of the cell processes with one another and the cell body play the decisive role in transmission, in which the coated pits may also be involved.

Hormonal imprinting at enzyme level

Each hormone represents simultaneously the ligand of the specific receptor, and the substrate of the enzyme responsible for its degradation or synthesis. Since, like receptors, the enzymes are molecules of a well-defined steric structure and have a ligand (the substrate), there is reason to postulate that imprinting also takes place at enzyme level. Six-week-old rats neonatally treated with DES or allylestrenol showed no change in basic hepatic mirosomal enzyme activity, but did show a change in response to testosterone, which failed to cause an activity increase relative to the control within 48 h. Allylestrenol proved to be more active than DES under the given conditions of the experiment[43].

These experimental observations suggest that *hormonal imprinting,* which had formerly been thought to be an exclusive property of molecules acting at receptor level, *can be regarded as a more general phenomenon, appearing and taking effect presumably in all binding site – ligand interrelationships, irrespective of the nature of the latter's function.*

The mechanism of imprinting

In the initial stage of evolution, as portrayed by present-day unicellular organisms, imprinting occurs in a dynamic open system, in which the continuously changing membrane patterns are screening the environment for specific signal molecules. In the presence of the specific signal molecule, the interacting membrane pattern becomes amplified and is transformed into a receptor. This persists if it acquires a selection advantage, through which the receptor structure, the cell carrying it, and the organism of which the cell forms a part, survive and become integrated into the mechanism of evolution[12, 15, 16]. In higher organisms the receptors are genetically encoded (pre-programed) structures, since at the level of multicellularity the cell has not the chance to form any kind of receptors for itself. However, the encoded receptor structure requires amplification, which is accomplished by imprinting.

The hormones act primarily on the membrane receptors or the cytosolic receptors, both in the critical perinatal period and later in life. The hormone molecules bound by membrane receptors become internalized into the cytoplasm, and if they do not become degraded therein, they presumably also develop action intracellularly[65, 81, 82, 88, 89]. Since the effect of hormonal imprinting is demonstrable in the progeny generation, there is reason to postulate that the hormone enters the cellular nucleus, too, and develops intranuclearly a gene level action. This presumption is supported by the experimental observation that the effect of hormonal imprinting was demonstrable in the F_1 offspring generation (fig. 10) of rats neonatally exposed to the hormone, although the circumstance that it did not reappear in the F_2-generation does not support suggestions of a strong genetic effect.

Radioisotope studies have shown that the labeled hormone content of lymphocytes was several orders greater in the perinatal period, and even at one week of age, than in adulthood[48]. Since the internalized hormones, at least the amino acid-type hormones, are also accumulated in the nucleus, their nuclear presence may contribute to the establishment of imprinting. Polypeptide hormones, too, have, apart from membrane receptors, intracellular and nuclear membrane associated receptors, which probably serve as mediators of imprinting[81, 82, 88, 89]. Since membrane DNA is being transposed to the nucleus, its involvement in the mechanism of imprinting cannot be excluded either.

Inhibitors of endocytosis and cellular protein synthesis inhibit hormone binding to cells in culture, and on simultaneous treatment with TSH they inhibit imprinting by TSH to different degrees, depending on TSH-binding[14]. Of the protein synthesis inhibitors, cycloheximide and cytochalasin (which acts on the microfilamentary system) inhibit imprinting to a still greater degree, from which it follows that hormonal imprinting presupposes the normal operation of the

94

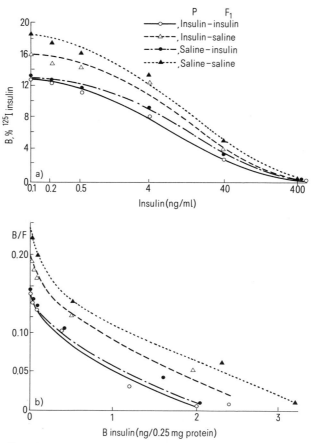

Figure 10. The effect of the neonatal hormone exposure of the parents (P) is still demonstrable in the offspring generation[28]. Hepatic insulin binding in female rats. Note the results of Scatchard analysis at bottom. The offspring- (F_1) of exposed parents showed a decreased binding relative to the control (given saline on both occasions) without being exposed themselves, a greater decrease after being exposed themselves when newborn, and this decrease was greater related to the insulin-exposed F_1 offspring of not exposed parents.

intracellular cxycloheximide-sensitive and cytochalasin-sensitive systems, irrespective of the binding relations. Although these findings suggest the involvement of receptor internalization and resysthesis in imprinting, the underlying mechanism is still obscure. Experiments in unicellular model systems are expected to throw more light on this problem.

Chemically the hormone receptors are glycoproteins, composed of a protein part and an oligosaccharide chain. Since the sugars also serve as determinant groups, they may play a certain role in hormonal imprinting. It appears that those lectins which bind to simple sugars, like for example pea lectin, are not able to induce imprinting either for themselves, or for hormones interacting with receptor structures which contain a like sugar molecule[18]. Against this Helix and Datura lectins, which bind to amino sugars, are able to induce imprinting for insulin, thus pretreatment with these lectins enhances insulin binding in culture. Datura lectin deserves special mention, for it imprints the receptor also for itself, to judge from its increased binding on reexposure. Presumably the hexosamine oligomeres (to which the Datura lectin binds) are also involved in the induction of imprinting.

The four stages of the development of encoded adaptation-requiring dynamic systems in mammals

It appears that, in any living being, adaptation (imprinting) can take place only at a given stage of ontogenetic development. This applies especially to mammals, for the chances of imprinting are limited by the particularities of the maternal-fetal relationship.

Experiments on steroid imprinting have shown that allylestrenol, which had a firm imprinting effect when applied in the perinatal or neonatal period, had no influence on the target organ during gestation, but developed a strong lifelong effect on the receptors of another organ[69]. Analysis of this phenomenon helps in an understanding of the nature of the factors averting or promoting imprinting, if the non-hormonal adaptation-requiring systems are also taken into consideration. Inhibition of prenatal steroid imprinting may have been associated partly with the protective barrier separating maternal from fetal circulation, and partly with the immaturity of the fetal steroid receptors. The time-dependent differences in action should therefore receive special consideration, and to facilitate this approach, we divided the ontogenetic development of the recognition system theoretically into four stages.

The first stage, which essentially coincides with the morphogenetic period, is characterized by full protection of the receptors against interaction with the hormone. At this stage presumably not so much the (as yet inaccessible) hormone receptors, as the receptors involved in cell-cell recognition (by marker recognition) are predominant, and contribute to the normal course of morphogenesis. Since little is known about the vulnerability of these receptors, it is poorly understood why destructive processes involving morphological changes also at organ level arise in that period. The marker-receptor system cannot be fully incriminated for such changes. At all events, its seems unlikely that either normal or abnormal imprinting could take place in the first stage of the recognition system's development.

The second stage is characterized by a relative protection of the developing hormone receptors, which have not yet come into their final shape (and quantity), because the factor(s) responsible for their amplification are lacking. Protection is justified, for maternal hormones are present in considerable amounts in the fetal circulation. Were adaptation (imprinting) to take place at that stage, it *would necessarily be oriented towards the quality and level of maternal hormones,* which would be deleterious, particularly in respect of sex differences. Relative protection, furnished by the immaturity, irresponsiveness, coating, etc.

of the developing receptor seems to prevent at this stage binding of the fetal hormone, or the like maternal hormone, to the receptors of the target organ. However, the factors furnishing relative protection of the target organ apparently cannot protect 'foreign' receptors against the influence of the hormone. Perhaps this mechanism was involved in the experiment mentioned earlier, in which perinatal allylestrenol treatment damaged severely the thymic glucocorticoid receptors of rats, but had no influence whatever on its own specific (uterine) receptors[69].

After the stage of relative protection follows the third stage, in which the system is open to lifelong hormonal imprinting, behavioral imprinting, and adjustment of the individual sexual constitution. This critical stage falls into the perinatal period. It cannot be precisely determined whether this stage commences before or after birth. Its beginning varies in all probability with the type of the developmental mechanism and also with the species, and may also depend on maturity at birth and on sex. Immunological adaptation (the capability of differentiating 'self' from 'non-self') begins somewhat earlier than hormonal and behavioral imprinting (although it is still in progress during their course). Behavioral imprinting can, naturally, begin only postnatally, since behavioral adaptation presupposes interaction with an external environment. The length of the open stage varies with the species, but it ends in all species in the early postnatal period. For example, the hormone receptors of the rat mature by the end of the third to fourth postnatal week.

The fourth stage can be defined as a state of relative openness, in which the learning mechanisms of the nervous system evolve, to persist for a lifetime. At this stage, probably, imprinting-like phenomena also occur in those cells which have preserved their fetal characteristics.

Similar phenomena take place in the regeneration of the adult liver, in which the mature cells undergo dedifferentiation (return of potential) and subsequently redifferentiation. Exposure to a hormone in a dedifferentiated stage will account for an imprinting-like effect, e.g. insulin treatment may alter the future hormone binding capacity[30], while allylestrenol or diethylstilbestrol treatment may enormously increase the phenobarbital-induced microsomal enzyme activity in females, but cause only a negligible decrease of it in males[44] (fig. 11). Imprinting may in all probability occur on the drug receptors of adults as well, to judge from the experimental observation that many new digoxine receptors appeared in the erythrocytes of humans after long-term treatment with digoxin[74]. This prompts the speculation that cells at least partly in possession of their embryonic potentials (the hemopoietic and gonadal cells) are available for imprinting also in adulthood, if their primary interaction with a foreign molecule acting at receptor level takes place in adult age.

Thus the dynamic recognition systems necessarily take their final shape by adaptation to the given (internal and/or external) environmental conditions. Adaptation may be qualitative or quantitative; for example, immune adaptation is almost exclusively qualitative, taking place in the very period in which 'self' and 'non-self' are differentiated for a lifetime. In the case of hormone receptors, adaptation has mainly quantitative aspects, since the future (final) receptor numbers also depend on the perinatally available hormone concentration. In other words, while the main function of hormonal imprinting is to

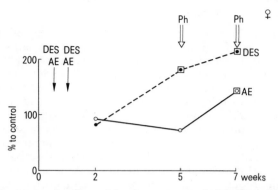

Figure 11. Hepatic microsomal enzyme activity after exposure to diethylstilbestrol (DES) or allylestrenol (AE) treatment during liver regeneration, with and without re-induction with phenobarbital (Ph). No appreciable effect was demonstrable without induction after two weeks, whereas induction with phenobarbital accounted for a significant activity increase in the DES-preexposed system, and a lesser, but still significant increase in the AE-preexposed system, relative to the controls not treated during the regeneration stage[44]. □, Significance $p < 0.01$. Indexes studied: cytochrome P_{450} (CO−Na dithionit), cytochrome P_{450}, cytochrome P_{450} metyrapone, anilin-p-hydroxylase, p-nitro-anisol-o-demethylase, p-nitro-phenol-hydroxylase.

increase the quantity of adequate receptors, immune adaptation furnishes a qualitative rather than a quantitative alteration of immune cells (receptors), probably because the immune system has to cope with the reception of both internal and external signals, whereas the hormone receptor is concerned with that of internal signals only. In this respect the immune system resembles the nervous system, being capable of 'learning' for a lifetime, despite the fact that it continues to refuse 'foreign' substances after the stage of self-recognition has terminated, unlike hormone receptors, whose 'learning' potential expires after acquiring the capacity of self-recognition.

Reference to immune-biological phenomena in the context of hormonal imprinting and ontogenesis seems to be justified, since many details of the two mechanisms seem to be alike. In the period of 'learning' to differentiate 'self' from 'non-self', the immune system suppresses the appropriate cell clones, whereas the hormone, on the contrary, activates the receptor system. Nevertheless, both mechanisms promote self-recognition. No 'self' input can take place in either system after termination of the critical adaptation period. Like immune receptors, the hormone receptors keep the difference between 'self' and 'foreign' in a lifelong 'memory'. A deeper insight into the mechanism of the immunological memory may probably throw more light on the mechanism of hormonal imprinting, and vice versa; better knowledge of the latter could perhaps promote the understanding of the immunological memory since, according to present knowledge, the genetic determinism of the immunological memory is also loose[60, 72], and perhaps requires the presence of the antigen for stabilization, exactly as receptor 'memory' requires the presence of the hormone in order to develop.

Medicinal aspects of hormonal imprinting

The quality of the hormone and of its target cell (receptor) is genetically determined. The time of appearance, and the quantitative relations of hormone and receptor are also genetically encoded. Disturbances can nevertheless occur in their interaction and mutual adaptation, e.g. if membrane differentiation occurs earlier, or the hormone appears later, than at the optimal time. Such circumstances may disturb receptor adaptation, and lead to an abnormal increase or decrease in cellular hormone binding capacity. If a structurally related (foreign) molecule appears in the critical period of primary receptor-hormone interaction, it may give rise to faulty imprinting by diminishing either the number or the affinity of receptors. The foreign molecules capable of affecting receptor development may not only be members of the same hormone family, but also certain products of the chemical and drug industry, which are structurally related to the specific hormone. Careful studies on steroid-like molecules, such as digoxin and benzpyrene, have attracted attention to the hazards of exposure to such molecules in the perinatal period. In the present era of 'chemicalization' the possible presence of such molecules in the air and water and/or in various drugs, may seriously affect the future health of the human infant if exposure occurs in the critical perinatal period.

Clarification of the still obscure details of the mechanism of hormonal imprinting may provide useful medical tools for the diagnostic identification, prevention, and/or correction of endocrine diseases, latent endocrinopathies, and non-physiological interventions hazardous to human health.

1   Alexander, R. W., Galper, J. B., Neer, E. J., and Smith, T. W., Non-co-ordinate development of β-adrenergic receptors and adenylate cyclase in chick heart. Biochem. J. *204* (1982) 825–830.
2   Amir, S. M., Sullivan, R. C., and Ingbar, S. H., Binding of bovine thyrotropin to receptor in rat testis and its interaction with gonadotropins. Endocrinology *103* (1978) 101–111.
3   Azuzizawa, M., Kutzman, G., Pekary, A. E., and Hershman, J. K., Comparison of the binding characteristics of bovine thyrotropin and human chorionic gonadotropin to thyroid plasma membrane. Endocrinology *101* (1977) 1880–1889.
4   Balázsi, I., Stützel, M., Varsányi-Nagy, M., and Karádi, I., Fat cell insulin receptors in children and adults. Diabetologia *15* (1978) 217.
5   Barraclough, C. A., Production of anoovulatory, sterile rats by single injection of testosterone propionate. Endocrinology *68* (1961) 61–67.
6   Barrington, E. J. W., Evolutionary aspects of hormonal structure and function, in: Comparative endocrinology. Eds P. J. Gaillard and H. Boer. Elsevier-North Holland, Amsterdam 1978.
7   Barrington, E. J. W., Hormones and evolution. Academic Press, London/New York 1979.
8   Bern, H. A., Gorski, R. A., and Kawashima, S., Long term effects of perinatal hormone administration. Science *181* (1973) 189–190.
9   Blazquez, E. B., Rubalcava, B., Montesano, R., Orci, L., and Unger, R. H., Development of insulin and glucagon binding and the adenylate cyclase response in liver membranes of the prenatal, postnatal and adult rat: evidence of glucagon 'resistance'. Endocrinology *98* (1976) 1014–1023.
10 Campbell, P. S., An early effect of testosterone propionate upon hypothalamic function in the neonatal rat. Experientia *39* (1983) 108–109.
11 Csaba, G., Phylogeny and ontogeny of hormone receptors: the selection theory of receptor formation and hormonal imprinting. Biol. Rev. *55* (1980) 47–63.
12 Csaba, G., Ontogeny and phylogeny of hormone receptors. Karger, Basel/New York 1981.

13 Csaba, G., The present state in the phylogeny and ontogeny of hormone receptors. Horm. Metab. Res. *16* (1984) 329–335.

14 Csaba, G., Hormone overlap, hormonal imprinting and receptor memory in tissue culture, in: Tissue Culture and Research. Eds P. Röhlich and E. Bácsy. Hung. Acad. Sci. Budapest 1984.

15 Csaba, G., The unicellular *Tetrahymena* as model cell for receptor research. Int. Rev. Cytol. *95* (1985) 327–377.

16 Csaba, G., Why do hormone receptors arise? Experientia *42* (1986) 715–718.

17 Csaba, G., and Bohdaneczky, E., Receptor development and hormone action. Effects of insulin and epinephrine on the glucagon content of newborn rat liver cultures. Acta physiol. hung. *58* (1981) 15–20.

18 Csaba, G., and Bohdaneczky, E., Induction of imprinting and 'memory' in Chang liver cells with lectin. Cell. molec. Biol. *30* (1984) 1–4.

19 Csaba, G., and Dobozy, O., The sensitivity of sugar receptor – analysis in adult animals of influences exerted at neonatal age. Endokrinologie *69* (1977) 227–232.

20 Csaba, G., Dobozy, O., and Kaizer, G., Study of FSH-TSH functional overlap by cockerel testicle. Horm. Metab. Res. *11* (1979) 689–692.

21 Csaba, G., Dobozy, O., and Kaizer, G., FSH-TSH functional overlap in cockerel testicle. Durable amplification of the hormone receptors by treatment at hatching. Horm. Metab. Res. *13* (1981) 177–179.

22 Csaba, G., Dobozy, O., and Deák, B. M., HCG-TSH overlap and induction of Galli-Mainini reaction with TSH in adult male frogs. Horm. Metab. Res. *14* (1982) 617–618.

23 Csaba, G., Dobozy, O., and Deák, B. M., Interaction of thyrotropin (TSH) and gonadotropins in the function of genital organs I. Acta physiol. hung. *61* (1983) 137–140.

24 Csaba, G., Dobozy, O., Shahin, M. A., and Dalló, J., Impact of a single neonatal gonadotropin (FSH+LH) or thyrotropin (TSH) treatment on the sexual behaviour of the adult male rat. Med. Biol. *62* (1984) 64–66.

25 Csaba, G., Dobozy, O., and Dalló, J., Influence of neonatal steroid (diethylstilbestrol, allylestrenol) treatment on the sexual behavior of the adult rat. Med. Biol. *64* (1986) 193–195.

26 Csaba, G., and Inczefi-Gonda, Á., Effect of benzo-a-pyrene treatment of neonatal and growing rats on steroid receptor binding capacity in adulthood. Gen. Pharmac. *15* (1984) 557–558.

27 Csaba, G., Inczefi-Gonda, Á., Dobozy, O., Varró, A., and Rablóczky G., Impact of neonatal treatment with cardioactive glycosides (digoxin, ouabain) on receptor binding capacity, blood level and cardiac function in the adult rat. Extension of the imprinting theory. Gen. Pharmac. *14* (1983) 709–711.

28 Csaba, G., Inczefi-Gonda, Á., and Dobozy, O., Hereditary transmission on the F_1 generation of hormonal imprinting (receptor memory) induced in rats by neonatal exposure to insulin. Acta physiol. hung. *63* (1984) 93–99.

29 Csaba, G., Inczefi-Gonda, Á., and Dobozy, O., Imprinting by steroids: a single neonatal treatment with diethylstilbestrol (DES) or allylestrenol (AE) gives rise to a lasting decrease in the number of rat uterine receptors. Acta physiol. hung. *67* (1986) 207–212.

30 Csaba, G., Inczefi-Gonda, Á., and Dobozy, O., Hormonal imprinting in adults. Insulin treatment during liver regeneration changes the later insulin binding capacity of rat liver insulin receptors. Exp. clin. Endocr. (1987) in press.

31 Csaba, G., Kovács, P., Török, O., Madarász, B., and Bohdaneczky, E., Cell-cell communication and hormonal imprinting. Transmission of receptor-level effect to cells, not directly influenced by the hormone. Cell. molec. Biol. (1987) in press.

32 Csaba, G., and Kovács, P., Development of hormonal imprinting by intercellular communication in Tetrahymena. Z. Naturforsch. (1987) in press.

33 Csaba, G., and Nagy, S. U., Plasticity of hormone receptors and possibility of their deformation in neonatal age. Experientia *32* (1976) 656–657.

34 Csaba, G., and Nagy, S. U., Influence of the neonatal suppression of TSH production (neonatal hyperthyroidism) on response to TSH in adulthood. J. Endocr. Invest. *8* (1985) 557–561.

35 Csaba, G., and Nagy, S. U., Influence of a single neonatal melatonin treatment on the basal and thyrotropin or melatonin modified blood thyroxine level of rats in adulthood. Acta physiol. hung., in press (1986).

36 Csaba, G., and Nagy, S. U., Can neonatal treatment with a related hormone adapt the receptor for itself? Acta physiol. hung. *67* (1986) 65–69.

37 Csaba, G., Németh, G., and Vargha, P., Development and persistence of receptor 'memory' in a unicellular model system. Expl Cell Biol. *52* (1982) 291–294.

100

38 Csaba, G., Rónai, A., László, V., Darvas, Zs., and Berzétei, I., Amplification of hormone receptors by neonatal oxytocin and vasopressin treatment. Horm. Metab. Res. *12* (1980) 28–31.

39 Csaba, G., Rónai, A., Dobozy, O., and Berzétei, I., Impact of neonatal catecholamine treatment on adult response to vasopressin and norepinephrine. Expl clin. Endocr. *84* (1984) 153–158.

40 Csaba, G., Shahin, M. A., and Dobozy, O., The overlapping effects of gonadotropins and TSH on embryonic chicken gonads. Archs Anat. Hist. Embryol. *63* (1980) 31-38.

41 Csaba, G., Sudár, F., and Dobozy, O., Triiodothyronine receptors in lymphocytes of newborn and adult rats. Horm. Metab. Res. *9* (1977) 499–501.

42 Csaba, G., and Sudár, F., Differentiation dependent alterations in lymphoytic triiodothyronine reception. Horm. Metab. Res. *10* (1978) 425–426.

43 Csaba, G., Szeberényi, Sz., and Dobozy, O., Influence of single neonatal treatment with allyle-strenol or diethylstilbestrol (DES) on microsomal enzyme activity in adulthood. Med. Biol. *64* (1986) 97–200.

44 Csaba, G., Szeberényi, Sz., and Dobozy, O., Hormonal imprinting of the microsomal enzyme systems in adults. Microsomal activity changes in response to steroid (DES, AE) treatment during liver regeneration. Horm. Metab. Res. (1987) in press.

45 Csaba, G., and Török, O., Impact of FSH-TSH overlap on the growth of Chinese hamster ovary (CHO) cell culture. Acta biol. hung. *34* (1983) 433–434.

46 Csaba, G., Török, O., and Kovács, P., Hormonal imprinting in cell culture I. Impact of single exposure to insulin on cellular insulin binding capacity in permanent cell lines. Acta physiol. hung. *64* (1984) 57–63.

47 Csaba, G., Török, O., and Kovács, P., Hormonal imprinting in cell culture II. Induction of hormonal imprinting and thyrotropin (TSH) – gonadotropin (FSH) overlap in a Chinese hamster ovary (CHO) cell line. Acta physiol. hung. *64* (1984) 135–138.

48 Csaba, G., and Ubornyák, L., Incorporation of hormones and hormone like materials by rat lymphocytes of different ontogenetic stages. Expl. clin. Endocr. *22* (1983) 68–72.

49 Dobozy, O., Balkányi, L., and Csaba, G., Thyroid cell hyporesponsiveness in cockerels treated with follicle stimulating hormone (FSH) or thyrotropin (TSH) at hatching. Horm. Metab. Res. *13* (1981) 587–588.

50 Döhler, K-D., Is female sexual differentiation hormone-mediated? TINS *1* (1978) 138–140.

51 Döhler, K-D., Srivastava, S. S., and Gorski, R., Postnatal tamoxifen treatment interferes with differentiation of the sexually dimorphic nucleus of the preoptic area in both male and female rats. Endocrinology *108* suppl. (1981) 187.

52 Döhler, K-D., Nordeen, E. J., and Yahr, P., Perinatal treatment of rats with an estrogen antagonist impairs estradiol uptake into brain cell nuclei in adulthood. Neurosci. Lett. suppl. *10* (1982) 149–150.

53 Dörner, G., Environment-dependent brain differentiation and fundamental processes of life. Acta biol. med. germ. *33* (1974) 129–148.

54 Dubois, J. D., and Dussault, J. H., Ontogenesis of thyroid function in the neonatal rat. Thyroxine (T_4) and triiodothyronine (T_3) production rates. Endocrinology *101* (1977) 438–439.

55 Dussault, J. H., Coulombre, P., and Walker, P., Effects of neonatal hyperthyroidism on the development of the hypothalamic-pituitary-thyroid axis in the rat. Endocrinology *110* (1982) 1037–1042.

56 Dussault, J. H., and Labrie, F., Development of the hypothalamic-pituitary-thyroid axis in the neonatal rat. Endocrinology *97* (1975) 1321–1324.

57 Fischer, D. A., and Klein, A. H., Thyroid development and disorders of thyroid function in the newborn. New Engl. J. Med. *304* (1981) 702–712.

58 Fletcher, W., and Greenan, J. R. T., Receptor mediated action without receptor occupancy. Endocrinology *116* (1985) 1660–1662.

59 Fontaine, Y-A., and Burzawa-Gerard, E., Esquisse de l'evolution des hormones gonadotropes et thyrotropes des vertebrés. Gen. comp. Endocr. *32* (1977) 341–347.

60 Fristrom, J. W., and Spieth, Ph., Principles of genetics. Blackwell, Oxford 1980.

61 Frowein, J., Engel, W., and Weise, H. C., HCG receptor present in the gonadotropin insensitive Leydig cell of the immature rat. Nature, New Biol. *246* (1973) 141–150.

62 Ganguli, S., Sirha, M., and Sperling, M. A., Ontogeny of insulin and glucagon receptors and the adenylate cyclase system in guinea pig liver. Pediatr. Res. *18* (1984) 558–565.

63 Ginsberg, B. H., Kahn, C. R., and Roth, J., The insulin receptor of the turkey erythrocyte: similarity to mammalian insulin receptors. Endocrinology *100* (1977) 520–525.

64 Gaspard, K. J., Klitgaard, H. M., and Wondergem, R., Sometomedin and thyroid hormones in the developing chick embryo. Proc. Soc. expl. Biol. Med. *166* (1981) 24–27.

65 Hollenberg, M. D., Receptor models and the action of neurotransmitters and hormones: some new perspectives, in: Neurotransmitter Receptor Binding. 2nd edn Ed. H. I. Yamamura. Raven Press, New York 1985.

66 Inczefi-Gonda, Á., Csaba, G., and Dobozy, O., Neonatal insulin treatment and adult receptor binding capacity in rats. Horm. Metab. Res. *14* (1982) 211-222.

67 Inczefi-Gonda, Á., Csaba, G., and Dobozy, O., Ouabain binding of the rat's heart muscle cells after neonatal glucocorticoid (triamcinolone) treatment. Acta physiol. hung. *67* (1986) 303–306.

68 Inczefi-Gonda, Á., and Csaba, G., Prolonged influence of a single neonatal steroid (dexamethasone) treatment on thymocytic steroid binding. Expl clin. Endocr. *85* (1985) 358–360.

69 Inczefi-Gonda, Á., Csaba, G., and Dobozy, O., Reduced thymic glucocorticoid reception in adult rats prenatally treated with allylestrenol. Acta physiol. hung. *67* (1986) 27–29.

70 Inczefi-Gonda, Á., Csaba, G., and Dobozy, O., Influence of a single neonatal treatment with steroid hormone or steroid like molecules on myocardial ouabain binding of the adult rat. Gen. Physiol. Biophys. (1986) in press.

71 King, D. B., King, C. R., and Eshleman, J. R., Serum triiodothyronine levels in the embryonic post hatching chicken with particular reference to feeding induced changes. Gen. comp. Endocr. *31* (1977) 216–219.

72 Leder, P., The genetics of antibody diversity. Scient. Am. *246* (1982) 72–83.

73 Maes, M., DeHertogh, R., Watrin-Granger, P., and Keterslegers, J. M., Ontogeny of liver somatotropic and lactogenic binding sites in male and female rats. Endocrinology *113* (1983) 1325–1332.

74 Malini, P. L., Strocki, E., Marata, A. M., and Ambrosini, E., Digitalis 'receptors' during chronic digoxin treatment . C. Expl Pharmac. Physiol. *11* (1984) 285–289.

75 Mess, B., and Strazniczky, K., Differentiation and function of the hypophyseal target organs system in chicken embryos. Hung. Acad. Sci., Budapest 1970.

76 Muggeo, M., Ginsberg, B. H., Roth, J., Neville, G. M., Meyts, P. de, and Kahn, C. R., The insulin receptor in vertebrates is functionally more conserved during evolution, than the insulin itself. Endocrinology *104* (1979) 1393–1402.

77 Muggeo, M., Obberghen, E. van, Kahn, C. R., Roth, J., Ginsberg, B. H., Meyts, P. de, Emdin, S. O., and Falkmer, S., The insulin receptor and insulin of the atlantic hagfish. Diabetes *28* (1979) 175–181.

78 Myachi, Y., Nieslag, E., and Lipsett, M. B., The secretion of gonadotropin and testosterone by the neonatal male rat. Endocrinology *91* (1973) 1-6.

79 Nagy, S. U., and Csaba, G., Long lasting amplification and deformation of thyroid receptors after thyrotropin (TSH) and gonadotropin (GTH) treatment of chickens in the foetal period. Acta physiol. hung. *56* (1980) 303–307.

80 Nagy, S. U., and Csaba, G., Dose dependence of the thyrotropin (TSH) receptor damaging effect of gonadotropin in the newborn rats. Acta physiol. hung. *56* (1980) 417–420.

81 Rao, C. V., Receptor for gonadotropins in human ovaries, in: Recent Advances in Fertility Research. Part A. pp. 123–135. Alan R. Liss, New York 1982.

82 Rao, C. V., and Chegini, N., Nuclear receptors for gonadotropins and prostaglandins, in: Evolution of Hormone Receptor Systems, pp. 413–423. Alan R. Liss, New York 1983.

83 Reichert, L. E., and Bhalla, V. K., Development of a radioligand tissue receptor assay for human follicle stimulating hormone. Endocrinology *94* (1974) 483–491.

84 Rónai, A. Z., Dobozy, O., Berzétei, I., Kurgyis, J., and Csaba, G., The effect of neonatal treatment of mice with opioid and dopaminergic agents on the late responsiveness of vasa deferentia to opioids in vitro. Acta biol. hung. *35* (1984) 43–47.

85 Sara, V. R., Hall, K., Misaki, M., Fryklund, L., Christensen, N., and Wetterberg, L., Ontogenesis of somatomedin and insulin receptors in the human fetus. J. clin. Invest. *71* (1983) 1084–1094.

86 Shahin, M. A., Török, O., and Csaba, G., The overlapping effects of thyrotropin and gonadotropin on chick embryo gonads in vitro. Acta morph. hung. *30* (1982) 109–125.

87 Shahin, M. A., Sudár, F., Dobozy, O., and Csaba, G., Electronmicroscopic study of the overlapping effect of thyrotropin and gonadotropins on the Leydig cells of 15-day-old chick embryo. Z. mikrosk.-anat. Forsch. *98* (1984) 926–938.

88 Szego, C. M., Parallels in the mode of action of peptide and steroid hormones: membrane effects and cellular entry, in: Structure and Function of Gonadotropins, pp. 431–472. Ed. K. W. McKerns. Plenum Press, New York 1978.

89 Szego, C. M., and Pietras, R. J., Lysosome function in cellular activation: propagation of the actions of hormones and other effectors. Int. Rev. Cytol. *88* (1984) 1–302.

90 Thorsson, A. V., and Hintz, R. L., Insulin receptors in the membrane increased in receptor affinity and number. New Engl. J. Med. *297* (1977) 908–912.

91 Ward, D. M., Correlation of hormonal structure with hormonal function in mammalian tissues, in: Invertebrate Endocrinology and Hormonal Heterophylly. Ed. W. J. Burdette. Springer, Berlin 1974.

92 Warren, D. W., Huhtaniemi, I. T., Topanainen, J., Dufau, M. L., and Catt, K. J., Ontogeny of gonadotropin receptors in the fetal and neonatal rat testis. Endocrinology *114* (1984) 470–476.

The special case of hormonal imprinting, the neonatal influence of sex

K. D. Döhler

Bissendorf Peptide GmbH, Burgwedeler Str. 25, D–3002 Wedemark 2 (Federal Republic of Germany)

The chain of events leading to reproductive success is based on the participation of a variety of organs and tissues with different structures and functions. The *brain* controls behavioral orientation and sexual identification as well as gonadotropic hormone (GTH) release from the pituitary gland. A sex specific pattern of gonadotropic hormone release will stimulate maturation and release of male or female germ cells respectively. A properly developed *internal duct system* will then transport the germ cells to the outside. Species with internal fertilization need appropriately developed *external genitalia* for the transfer of germ cells from the male to the female individual.

Historical perspectives

The question about which factors may determine the fate of a developing fetus, causing it to become either male or female, has occupied many previous cultures and scientists. Ancient Greek ideas about sexual differentiation centered mainly around two hypotheses. The 'hypothesis of laterality', established by Anaxagoras of Clazomenae (about 440 BC), claimed that semen from the right testis would produce male offspring, semen from the left testis would produce females. This hypothesis further claimed that male fetuses are carried in the right horn of the uterus, females in the left[1]. The 'thermal hypothesis' of Empedokles of Akras (about 460 BC) claimed that temperature was an important factor in sex determination[85]. Conception in a hot uterus would produce a male, in a cold uterus a female. Aristotle of Stagirus (384 to 322 BC) favored the thermal hypothesis. He observed in sheep and goats that they would produce male offspring when warm winds were blowing from the south during copulation, but female offspring when cold winds were blowing from the north[4]. Plato postulated that the first human generation consisted only of men. Those men of the first generation, who had been cowardly or had spent most of their lives in wrong-doing, were reborn in the second generation as women[86].

Environmental influences on sexual differentiation

The 'thermal hypothesis' of Empedokles may actually not be that far off after all. It has been shown that frog larvae develop a male phenotype when raised at an elevated water temperature; at a low temperature they develop into females[83]. In some species of lizards breeding of the eggs at temperatures below 26 °C will prime the embryos for female development, whereas at temperatures above 26°C the embryos will develop into males[95]. In two species of turtles, *Emys orbicularis* and *Testudo graeca* the temperature effect on sexual differentiation is reversed. Male development is induced during breeding at temperatures below 28°C and female development is induced during breeding at above 32°C[82].

Another environmental influence which may effect sexual differentiation is the concentration of potassium and calcium ions in the water. Three- to four-fold elevation of calcium ions in the water will stimulate the larvae of *Discoglossus pictus* to develop into females. Five- to six-fold elevation of calcium ions will stimulate the same larvae to develop into males[98].

Genetic influences on sexual differentiation

After the ancient times of the early Greek philosophers it took almost 2500 years before the role of the Y-chromosome in masculinization of the gonads was discovered. In mammals, a gene on the male Y-chromosome stimulates the production of a cell surface antigen, the histocompatibility-Y-antigen (H-Y-antigen). Under the influence of H-Y-antigen the gonads will differentiate into testes. Lack of H-Y-antigen, as in the female, will result in differentiation of ovaries. Although differentiation of the mammalian gonads is under chromosomal control, differentiation of other sexual structures (reproductive tract, external genitalia and even the brain) is now known to be controlled by an imprinting action of hormones during fetal or neonatal life.

Hormonal influences on sexual differentiation

1. Differentiation of the gonads. Differentiation of the mammalian gonads is under chromosomal control and cannot be influenced by the action of hormones. Although in lower vertebrates gonadal differentiation may also be primarily under genetic control, it was shown in several species of fish and amphibians that early hormonal influences may sex-reverse the differentiation of the gonads without sex-reversing the genetic information. When female larvae of the medaka fish *Oryzias latipes* are treated with androgens, they will develop into fully reproductive males. These males will produce only female offspring after breeding with normal females. When male larvae of *Oryzias latipes* are treated with estrogens, they will develop into fully reproductive females. During breeding with normal males these females (XY genotype) will produce 25% female offspring (XX genotype) and 75% male (XY and YY genotype) offspring[114-116]. Gonadal sex reversal was also observed in several amphibian species when the larvae were raised in water which contained estrogenic or androgenic hormones[14, 15].

2. Differentiation of reproductive tract and external genitalia. In mammals, differentiation of the gonads cannot be influenced by hormones, but differentiation of the reproductive tract and the external genitalia depends exclusively on the imprinting (priming) action of gonadal hormones. The *sertoli* cells in the testes produce a locally acting substance, the Müllerian inhibiting factor. This factor, the structure of which is still unknown, causes regression of the female embryonic reproductive tract, the Müllerian ducts (fig. 1). Without the priming action of this factor, which does not exist in females, the Müllerian ducts will develop into fallopian tubes, uterus and upper vagina.

The *leydig cells* in the testes produce testosterone, a male sex hormone (androgen). Under the priming influence of testosterone the male embryonic reproductive tract, the Wolffian ducts, will develop into vasa deferentia, seminal vesicles and epididymis (fig. 1). Differentiation of the external male sex organs (penis, scrotum and sinus urogenitalis) occurs under the priming influence of another androgen, 5α-dihydrotestosterone (DHT). In the target cells this androgen is converted from testosterone by the enzyme 5α-reductase. DHT is subsequently bound by intracellular cytosolic androgen receptors and this receptor-hormone complex is then translocated into the cell nucleus. Interaction of the receptor-hormone complex with the genetic material of the nucleus stimulates sexual differentiation of the respective organ irreversibly. The exact molecular mechanism of this irreversible imprinting action is still unknown. In cases where DHT is absent during the sensitive developmental phase – this is normally the case in females, but also in males with deficiency in 5α-reductase[79] – the external genitalia will develop in the female direction (vulva and lower vagina). The female type of genital development also occurs when the genital tissue is non-responsive to androgens, owing to an androgen receptor defect, as it is in the syndrome of testicular feminization[5, 71].

3. Differentiation of the brain. Sexually dimorphic brain functions. The most obvious functional differences between male and female animals are those involved in reproductive physiology and reproductive behavior. The best-studied animal model in this respect is the rat. In the female rat, rising plasma titers of estrogens trigger a cyclic neural stimulus which activates the release of gonadotropin-releasing hormone(s) (GnRH) from the hypothalamus (positive

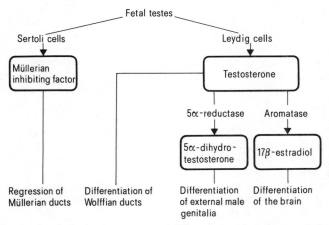

Figure 1. Influence of testicular hormones on sexual differentiation of sex ducts, genitalia and brain.

feedback). GnRH, in turn, stimulates the release of luteinizing hormone (LH) and follicle stimulating hormone (FSH) from the pituitary gland. The gonado-tropins FSH and LH stimulate follicular maturation in the ovaries and trigger ovulation. In the male rat, rising plasma titers either of estrogens or of andro-gens are unable to stimulate the release of GnRH. The neural substrate which controls GnRH release has apparently developed differently in males and females.

The neural substrate which controls sexual behavior has also developed along different lines in males and females. Under the influence of estrogens and progesterone, adult female rats will respond to the mounting attempts of a sexually active male by an arching of the back, the so-called lordosis reflex. Adult male rats will hardly show any lordosis behavior, even if given the same hormone treatment. Under the influence of testosterone, adult male rats will show vigorous mounting, intromission and ejaculatory behavior towards a receptive female, whereas female rats will show little or no such response when treated with testosterone.

Organization of sexually dimorphic brain functions. Sexual differentiation of the brain has been studied most thoroughly in the rat. In 1936 Pfeiffer[81] presented evidence that there is a critical period during early postnatal development of the rat, during which differentiation of the pattern of anterior pituitary hormone secretion can be influenced permanently by testicular hormone action. He removed the testes of newborn male rats and replaced them with ovaries when the animals were adult. These male animals showed the female capacity to form corpora lutea in the grafted ovarian tissue. Newborn females, implanted with testes from littermate males, did not show estrous cycles or form corpora lutea in their ovaries when adult. Pfeiffer[81] had originally concluded that the pituitary gland is sexually dimorphic in function, but it was shown later that hormone release from the pituitary gland and, thus, ovulation is under the control of the central nervous system[42].

Present knowledge of hormonal influences on the development of sexually dimorphic brain functions is based on a great number of studies, most of which have been carried out in the last 25 years. The individual contributions to the field of sexual brain differentiation have been discussed in several excellent reviews[13, 37, 50, 51, 84]. In summary, there is a sensitive developmental period during which sexual differentiation of neural substrates proceeds irreversibly under the influence of gonadal hormones. In the rat this period starts a few days before birth and ends approximately 10 days after birth. Female rats, treated during this sensitive period with testosterone or estradiol, will permanently lose the capacity to release GnRH in response to estrogenic stimulation, and will lose the capacity to show lordosis behavior ('defeminization'). Instead, they will develop the capacity to show the complete masculine sexual behavior pattern following administration of testosterone in adulthood ('masculinization'). If castrated perinatally, male rats become unable to display male sexual behavior patterns after treatment with testosterone in adulthood ('demasculinization'). Instead, they will develop the capacity to show lordosis behavior, and to respond in adulthood with a positive GnRH feedback to estrogen treatment ('feminization').

These studies indicate that androgens and/or estrogens, whether released by the testes or applied exogenously during the perinatal period, will permanently defeminize and masculinize neural substrates controlling sexually dimorphic brain functions (fig. 2).

Sexually dimorphic brain structures. Despite the well-known sex differences in brain functions, brain structure was for a long time believed to be essentially the same in males and females. The first anatomical sex differences observed in the mammalian brain were rather subtle. In rats, Pfaff[80] as well as Dörner and Staudt[38, 39] observed differences between the sexes in the size of nerve cell nuclei. Sex-linked differences in the pattern of neuronal connections were observed in rodents by Raisman and Field[87], Dyer et al.[40], Dyer[41], Greenough et al.[52], Nishizuka and Arai[76] and by De Vries et al.[21, 22]. In all these cases the sex differences proved to be dependent upon the degree of androgen exposure during the perinatal period.

The first discovery of a gross sexual dimorphism of the brain was made by Nottebohm and Arnold[77] on two species of song birds. During a reinvestigation of the male and female rat brain, Gorski et al.[49] observed a striking sexual dimorphism in the gross morphology of the medial preoptic area. The volume of an intensely staining area, now called the sexually dimorphic nucleus of the preoptic area (SDN-POA), is several times larger in adult male rats than in females[47–49]. The development of this nucleus starts during late fetal life[54, 56, 57] and depends on the hormonal environment during the critical period of sexual differentiation[24–28, 32, 34, 49, 55].

The sexually dimorphic nucleus of the preoptic area (SDN-POA): development and differentiation. Development of the SDN-POA was shown to start during late fetal life and to extend throughout the first ten days of postnatal life[54, 56, 57]. This developmental period is identical with the period when sexual differentiation of brain function proceeds under the influence of gonadal hormones. A series of studies was performed during recent years in order to test the influence of hormones perinatally on development and differentiation of the SDN-POA. Neonatal castration of male rats reduced the volume of the SDN-POA permanently[49, 55]. Reimplantation of a testis or treatment with a single

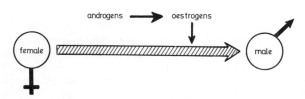

Figure 2. Diagram summarizing contemporary thinking on the mechanism of sexual brain differentiation. The genetic program for brain development is thought to be inherently female. It will remain female unless male differentiation tendencies are epigenetically triggered by androgens or estrogens during a sensitive period. The organizational effects of androgens are thought to be mediated by intracellular conversion of these hormones in certain brain areas to estrogens (aromatization hypothesis). Diagram from Döhler and Hancke[29] after modification.

injection of testosterone propionate (TP) one day after neonatal castration restored SDN-POA volume in male rats to normal[55]. Treatment of female rats with a single injection of TP either prenatally (F. C. Davis, unpublished observations) or postnatally[49, 55] increased SDN-POA volume significantly; however, the volume of the SDN-POA in these animals was still significantly smaller than that in normal male rats. Only the extended pre- and postnatal treatment of female rats with TP resulted in SDN-POA differentiation equivalent to that of normal males[24, 26]. The treatment of male rats pre- and postnatally with TP did not increase the size of their SDN-POA above normal[24, 26].

A closer look at female sexual differentiation of the brain. Sexual organization of the brain is thought to be inherently female unless male differentiation is superimposed by androgens or estrogens during a critical period of development. The organizational effects of androgens are thought to be mediated by intracellular conversion of these hormones in certain brain areas to estrogens. In other words, female differentiation is thought to proceed in the absence of specific hormonal influences, whereas male differentiation requires estrogenic stimulation (fig. 2).

The assumption that female sexual differentiation proceeds normally in the absence of gonadal hormones is based upon the early observation by Jost[60] that gonadectomy of female rabbit fetuses does not interfere with female differentiation. Estrogen concentrations in mammalian fetuses are known to be very high, often higher than during later reproductive live[9, 89, 90, 99, 110, 111], and fetal ovariectomy was assumed to clear the fetal blood circulation of estrogens. Recently it was shown, however, that the fetal ovaries are in fact not the major source of the estrogens found in the fetal circulation[45, 63]. In the best-studied species, the human, the primary source of estrogens during pregnancy are the fetal and, to a lesser degree, the maternal adrenals. The adrenals secrete aromatizable androgens (mainly dehydroepiandrosterone sulfate) which are aromatized to estrogens in the placenta[63]. There is, therefore, no reason to believe that fetal gonadectomy renders the fetus free from estrogens.

In the rat, the determination of sexually dimorphic brain differentiation occurs mostly after birth. The assumption that female sexual differentiation of the rat brain would proceed in the absence of gonadal hormones is based on the early observation by Pfeiffer[81], that ovariectomy of newborn female rats did not interfere with female differentiation of the brain.

The role of alpha-fetoproteins. Meanwhile we know very well that postnatal ovariectomy of rats does not clear the blood circulation of estrogenic hormones[113]. This is due to the presence of high levels of estrogen-binding alpha-fetoproteins (AFP)[78, 88] which protect circulating estrogens from metabolism. In fact it was shown that the ovaries of newborn rats do not secrete any estrogens before day 7 of life[64], and it seems most likely that the high levels of estrogens observed in newborn male and female rats[36] are actually remainders of maternal/placental origin from prenatal life (for discussion see ref. 31).

The biological role of AFP during the fetal and neonatal period is rather speculative. It was originally assumed that AFP may prevent circulating estrogens from interacting with the developing brain, thus protecting it from masculinization[68]. This assumption becomes highly dubious, however, in view of the intra-neuronal localization of AFP[10, 104]. It should also be considered that without AFP there would not be any estrogens in the blood circulation of postnatal female rats, since the ovaries do not release estrogens during the first week of life[64].

Recent investigations favor the proposition that the biological purpose of AFP may actually be to protect estrogens from enzymatic degradation and to inhibit estrogen uptake by the liver, thus precluding metabolism and excretion (see refs. 31 and 105 for reviews). AFP may even act as carrier for the transport of estrogens into brain cells, as suggested by Döhler[23] and by Toran-Allerand[105]. The overall result would be the conservation of vital estrogens, which are crucial not only for female differentiation of the brain, but also for certain aspects of brain development.

The necessity for perinatal imprinting by estrogenic hormones. Studies of the hormonal influence on female differentiation of the brain have been hampered by the fact that gonadal/placental hormones cannot effectively be removed from the blood circulation of fetal mammals and of postnatal rats and mice. In a series of studies, therefore, Döhler and co-workers[27, 28, 31, 33, 34, 53, 109] adopted the approach of inactivating the endogenous estrogens by treating newborn female rats with the estrogen antagonists tamoxifen or LY 117018 respectively. Both estrogen antagonists inhibit the biological effects of estrogens by competing with estrogens for intracellular estrogen receptor binding sites[58, 59, 75, 101].

Postnatal treatment with tamoxifen[31, 34, 53] or with LY 117018 (Ganzemüller, Veit and Döhler, unpublished) inhibited permanently the differentiation of a positive feedback mechanism for the estrogen-stimulated release of luteinizing hormone (LH), and it inhibited differentiation of the capacity to show female sexual receptivity. These results confirm and extend studies which had been performed with other estrogen antagonists[16, 44, 69]. Since female sexual differentiation of the brain is thought to occur in the complete absence of hormones, the dramatic effects of estrogen antagonists on the developing female brain seem surprising.

On the other hand, the initial steps of the intracellular action of estrogens and estrogen antagonists are similar to some extent. Tamoxifen and LY 117018 are known to bind to intracellular estrogen receptors in an apparently similar fashion as is done by estradiol and other estrogens[58, 59, 75, 101]. On the basis of such similarities one is tempted to consider the possibility that the defeminizing effects of tamoxifen on the developing female brain may actually be due to estrogenic effects rather than to estrogen antagonism. The available results indicate, however, that the permanent biological effects, induced by postnatal treatment of female rats with tamoxifen or LY 117018, are different from the permanent biological effects which have previously been shown to occur after postnatal treatment of female rats with estrogens[13, 26–28, 32, 34, 84]. Although tamoxifen and LY 117018 inhibited differentiation of female sexual behavior patterns, they did not stimulate the organization of male sexual behavior patterns[53], an event which is known to be stimulated by postnatal action of estrogens[13, 84].

The conclusion that the two estrogen antagonists did not act like estrogens, but instead prevented the activity of estrogens postnatally, is supported by the finding that development and differentiation of the SDN-POA is stimulated by perinatal treatment with an estrogen[26, 28, 32], but inhibited by similar treatment with tamoxifen[27, 28, 34]. Furthermore, the defeminizing effect of tamoxifen on organization of female sexual brain functions was actually attenuated by concomitant treatment with estradiol[31, 53].

The conclusion that female sexual differentiation of the brain may not proceed without hormones, but may need estrogenic stimulation, is supported by the results from several other studies. Toran-Allerand[103, 105] demonstrated that hypothalamic neurons of newborn mice do not develop neurite processes in vitro when the culture medium is devoid of estrogens. Vom Saal et al.[110, 111] observed that female mice which were located in utero between two other females had higher levels of estradiol in their amnionic fluid, and showed better adult sexual performance, than did their female litter mates which had been located in utero between two male fetuses.

In summary, the available data[10, 16, 27, 28, 31, 33, 34, 44, 53, 69, 103–105, 109–111] suggest that female sexual differentiation of the brain, or even brain development per se, may require perinatal estrogenic stimulation for its full expression. Therefore, the capacity for the normal display of female sexual behavior and for the cyclic release of gonadotropins is not, as has been assumed, inherent to central nervous tissue, but depends on active hormonal induction during a sensitive period of development. Perinatal antagonism of estrogenic activity thus produces animals which are neither male nor female, behaviorally and physiologically speaking. In adulthood they respond neither to estradiol nor to testosterone. Requirements for estrogenic influences on male and female brain differentiation, functional and structural, may be quantitative rather than qualitative[23, 29, 31].

A closer look at male sexual differentiation of the brain. Male sexual differentiation is controlled by testicular hormon. Previous observations, indicating that non aromatizable androgens are not able to stimulate masculinization or defeminization of brain functions, whereas estrogens or aromatizable androgens are quite effective in this regard, generated the hypothesis that male differentiation of the brain is exclusively estrogen dependent. Androgens are thought to be active in brain differentiation only after being enzymatically aromatized to estrogens (fig. 2). The following paragraphs will indicate, however, that the processes of masculinization or defeminization of different brain structures and functions are under the control of more complex hormonal mechanisms.

Structural development and differentiation depends on estrogens. Although pre- and postnatal treatment of rats with TP was shown to substitute fully for testicular activities in stimulating SDN-POA development[24, 26], the prime candidates for the control of SDN-POA differentiation do not seem to be androgens as such, but rather estrogens. This conclusion is supported by several observations: 1) Female rats which had been treated pre- and postnatally with the synthetic estrogen diethylstilbestrol developed a significantly enlarged SDN-POA, which was similar in volume to that of control males[26, 28, 32]. This observation indicates that estrogens can stimulate SDN-POA development directly. The treatment of male rats pre- and postnatally with diethylstilbestrol did not

increase the size of their SDN-POA above normal[26, 32]. 2) Male rats, treated pre-
and postnatally with the androgen antagonist cyproterone acetate, developed
female genitalia, but the volume of their SDN-POA was not reduced[27, 28]. 3)
Male rats, treated pre- and postnatally with the estrogen antagonist tamoxifen,
developed male genitalia, but the volume of their SDN-POA was significantly
reduced and was similar to that of control female rats[27, 28]. The normal development of male genitalia in these animals and the observation that pre- and
postnatal treatment of male rats with tamoxifen did not influence serum levels
of testosterone[27], indicate that tamoxifen did not act via inhibition of testosterone release from the testes. Instead, the growth inhibiting influence of the
estrogen antagonist on the SDN-POA, a brain area with known sensitivity to
estrogens[26, 32, 48, 100], seems most likely to be due to local interference with the
activity of estrogens, which may have derived via enzymatic conversion from
circulating androgens.

In the adult organism tamoxifen is known to bind to intracellular estrogen
receptors and to prevent estrogen uptake as it inhibits cytosol receptor replen-
ishment[58, 59, 75]. Tamoxifen may act similarly in the developing organism. After
aromatization of testicular androgens into estrogens tamoxifen may have interfered with estrogen uptake into cell nuclei of the SDN-POA by occupying
intracellular estrogen receptors. The inhibitory effect of pre- and postnatal
tamoxifen on growth and differentiation of the SDN-POA in male rats indicates
that not only functional, but also structural differentiation of the male rat brain
may be dependent on aromatization of testicular androgens into estrogens and
the subsequent interaction of these estrogens with the nuclear material.

The observation that the androgen antagonist cyproterone acetate did not
interfere with growth and differentiation of the SDN-POA indicates that androgens are not the primary stimulators of SDN-POA differentiation. Androgens
seem to be the substrate, which has to be converted into estrogens before being
able to activate SDN-POA differentiation.

Functional differentiation: estrogens alone are without effect. The fact that
cyproterone acetate did not inhibit development of the SDN-POA indicates
further that cyproterone acetate did not interfere with androgen entry into
preoptic-hypothalamic nerve cells or with aromatization of androgens into
estrogens. Aromatization of androgens into estrogens is generally considered to
be a prerequisite for preoptic-hypothalamic masculinization and defeminization of brain functions. Pre- and postnatal treatment of rats with cyproterone
acetate was repeatedly shown to feminize permanently sexual behavior patterns
and the mode of gonadotropin release in males[73, 74], and to inhibit the defeminizing action of exogenous testosterone in females[3].

Since differentiation of the SDN-POA was more complete after treatment of
male rats perinatally with cyproterone acetate[27] than after treatment of female
rats postnatally with a single, anovulation-inducing dose of testosterone propionate[49, 55], estrogenic interaction with preoptic-hypothalamic tissue must have
been more intense (and/or more physiological?) during treatment with cyproterone acetate than during treatment with testosterone propionate. Nevertheless, the intensive hypothalamic interaction with estrogens during cyproterone acetate treatment did not disrupt differentiation of the cyclic mode of
gonadotropin release or differentiation of female sexual behavior patterns[46, 73, 74],

whereas the less intensive (or less physiological?) estrogenic interaction during postnatal treatment with TP interfered with both events.

Thus the question arises, how exactly does cyproterone acetate interfere with sexual differentiation of brain function?

Functional differentiation; the necessity for androgens. The observation that only aromatizable androgens stimulate differentiation of male sexual behavior patterns in female rats, and that non-aromatizable androgens seem to be without this capacity (for reviews see refs. 13, 51, 84), and the observation that masculinization of sexual behavior patterns can be inhibited by postnatal treatment of rats with estrogen antagonists[11, 69, 97, 109], seem to indicate that estrogens are the major effective hormones which stimulate the differentiation of male sexual behavior patterns. However, this conclusion does not fit with the observation that perinatal treatment of male rats with an aromatization inhibitor did not inhibit differentiation of the capacity of male sexual behavior[112], whereas perinatal treatment with an androgen antagonist inhibited differentiaton of the capacity for male sexual behavior, despite its inability to prevent aromatization of testicular androgens into estrogens[27, 28]. The latter observations suggest that estrogens per se may be less important and androgens per se may be more important for the organization of male sexual brain functions, than has been assumed.

Testosterone is known to enter androgen target cells in the brain. Within the target cells testosterone is subjected to aromatization or to 5α-reduction, the principal metabolites being estradiol and 5α-dihydrotestosterone (DHT). Both hormones are bound with high affinity to specific cytoplasmic receptor proteins and are then translocated into the cell nucleus where they stimulate a characteristic biological response. Cyproterone acetate does not prevent androgen entry into hypothalamic cells, nor does it influence androgen metabolism[67, 102]. Its main antagonistic activity seems to be based on interference with intracellular androgen binding to specific androgen receptors in the cytosol and prevention of the translocation of the receptor-androgen complex into the cell nucleus[67, 102]. Thus, the activity of cyproterone acetate is directed against androgen-mediated events, but is not directed against estrogen-mediated events.

With reference to sexual differentiation of the brain, this discussion points to one necessary conclusion: masculinization and defeminization of sexual brain functions in the rat seem to be mediated not only by estrogens alone, but also seem to require the participation of androgens per se. Androgenic and estrogenic components seem to be required for complete masculinization and defeminization of sexual brain functions. Interference of hormone antagonists with one or the other component results in incomplete organization of the male brain (fig. 3).

Metabolism of non-aromatizable androgens is very rapid in the rat, but slow in the rhesus monkey[51]. The fact that non-aromatizable androgens are quite effective in stimulating differentiation of male sexual behavior patterns in monkeys[51, 84], but not in rats, may be a result of the different speed of metabolism, rather than an indication of a different mechanism of brain differentiation. The inefficiency of DHT, in contrast to testosterone propionate, in stimulating the male type of genital differentiation in female rats, when given at high daily doses during the last week of fetal life[27, 28], is a strong indication that the non-aromatizable androgen never reaches the target organ.

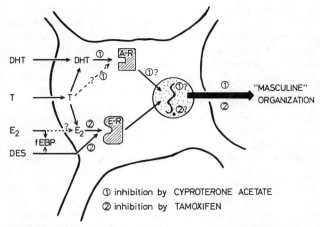

Figure 3. Schematic representation of the androgenic and estrogenic hormone environment with plasma and neuronal compartments in the perinatal rat in relation to masculine organization of the brain. The diagram indicates that interference with androgen action by the androgen antagonist cyproterone acetate (1) or interference with estrogen action by the estrogen antagonist tamoxifen (2) will interfere with masculine organization of the brain. The possible steps of interference are listed A-R = cytoplasmic androgen receptor; DES, diethylstilbestrol; DHT, 5α-dihydrotestosterone; E_2, estradiol; E-R, cytoplasmic estrogen receptor; fEBP, fetal estrogen-binding protein; T, testosterone.

In conclusion, during differentiation of male sexual brain function estrogens may be supportive to the primary action of androgens, rather than directive.

Hormonal imprinting and hormone receptors. The intracellular and molecular mechanisms of hormonal imprinting during the process of sexual differentiation are poorly understood. We know that steroid hormones stimulate neurite outgrowth in the central nervous system[103, 105], a process which may result in the development of sex specific neural circuitries. Whether this influence of steroids may be mediated via effects on the plasma membrane, or via effects on the genetic code, receptor-mediated or not receptor-mediated, is still a matter of speculation. It is a fact that development of the central nervous system and programing of neural circuitries takes place in parallel with dramatic changes in steroid hormone receptor levels in various target tissues (see ref. 66 for review), and with changes in serum concentrations of estrogen binding proteins[78, 88]. All of these events occur against a background of greatly elevated plasma estrogen concentrations in both sexes[36], elevated plasma androgen levels in males[36], and dramatic increases in serum levels of triiodothyronine and thyroxine in both sexes[35].

It is reasonable to assume that steroid receptors are involved in the process of steroid-induced sexual differentiation, since the syndrome of testicular feminization is due to a defect in androgen receptor development[5, 70]. Individuals with testicular feminization are genetically males, but since their tissues are unresponsive to androgens they do not develop male genitalia or male internal sex ducts. Now the question arises, which mechanism stimulates induction and development of reproductive hormone receptor systems? The answer is, we do not really know. Since, however, the syndrome of testicular feminization is originally a genetic defect, it can be concluded that the genetic code participates in some way in the formation of androgen receptors. The absence of known genetic defects of estrogen receptors suggests that such defects would be incompatible with life and underlines the possible fundamental importance of estrogen receptors for the development and differentiation of the central nervous system.

Since the genetic code is identical in each cell of one organism, and since steroid receptor development does not occur in each cell of the body, but is restricted to cells in specific steroid target tissues, it can be concluded, that the development of steroid receptors is not due exclusively to genomic influences. Instead, activation of genomic information must depend on particular intracellular environments. It was mentioned previously that in lower vertebrates the hormonal environment can totally override the sex-determining genetic mechanism, even in germ cells, without altering the genetic code[116]. These observations indicate that, in regard to sex-determining mechanisms, the genetic code does not provide the final anatomical or physiological substrate. Instead, the genetic code only provides 'precursor information' which, during early development of an individual organism, will be shaped (imprinted) into the final substrate by hormones.

It has been shown that estrogen receptors in cells of the adult uterus are present in different configurations[65]. The original receptor, synthesized in the endoplasmatic reticulum, has been described as being of precursor nature[65]. This receptor is modified at least twice before it is released into the cytosol[65]. In the absence of estradiol, the original cytosol receptor is compounded in some unknown fashion into a storage form, which can be recalled at any time. In the presence of estradiol the cytosol receptor dimerizes into the active form and translocates into the nucleus[65]. These data indicate that, even in the adult organism, it is not the active receptor which is synthesized as a result of genetic information, but rather a precursor form of the receptor. The final shaping of the receptor is performed by the cellular (hormonal) environment.

Ontogeny of estrogen receptors. In prenatal rats estrogen receptors are still undetectable, but they increase in number substantially in hypothalamus, amygdala and cortex during the postnatal period. Development and maturation of estrogen receptor systems involves two well-defined phases. The initial phase, extending up to postnatal day 5, is characterized by a rapid increase in receptor concentrations throughout the brain, from undetectable levels 2 to 4 days before birth. The second phase, which occurs between postnatal days 5 and 25, transforms the neonatal receptor distribution into the adult pattern through a series of dynamic changes in estrogen receptor levels in different brain regions. In the cerebral cortex, receptor concentrations first increase to a peak at day 10 and then decline rapidly between days 10 and 15. Hypothalamic and pituitary receptor levels also increase up to day 10, but decline only slightly between days 10 and 25. In the preoptic area estrogen receptor levels increase steadily throughout the first 25 days of life; in the amygdala and midbrain they do not change significantly from post natal day 3 onward (see ref. 66 for review).

Alpha-fetoprotein, a precursor for estrogen receptors? The fact that estrogen receptors are as yet undetectable prenatally does not exclude the possibility that they may actually be present in some precursor-form. Good candidates for such precursors would be the different molecular forms of alpha-fetoproteins, which have been detected in serum[78, 88, 108] and in uterus[7, 106, 107] and brain cells[6, 100, 104, 105]. It was demonstrated by Vallette et al.[108] that high and low carbohydrate forms of alpha-fetoprotein exist in the serum of developing rats. The different forms of alpha-fetoprotein have estrogen-binding affinities, which range from 10^{-7} to 10^{-9} M/l. The estrogen-binding affinity of intracellular estrogen receptors is one order of magnitude higher (10^{-10} M/l) than the highest affinity of alpha-feto-protein[2]. Uriel[106] demonstrated that in the uterus of immature rats in intracellular cytosol receptor was transformed by 0.4 M KCl into an estrogen-binding protein, whose antigenicity, sedimentation coefficient and binding specificity resembled closely that of alpha-fetoprotein. These data were confirmed by some groups[8, 96], but not by others[6]. Sometimes methodological problems may mask the true results, and if something cannot be seen with the methodology used, it does not necessarily mean that it does not exist.

A controversy of similar nature was raised by Toran-Allerand[104], who observed intracellular location of immunologically active alpha-fetoprotein in neuronal groups throughout the developing rodent brain of both sexes, with the exception of certain areas, which are known to contain high affinity estrogen receptor sites and to be specific targets for estrogens. The significance of this absence of immunoreactive alpha-fetoprotein within the presumed target regions for estrogens during the critical period for sexual differentiation is not known. It is conceivable that estrogen-sensitive regions may be 'protected' by the specific inability of their neurons to take up alpha-fetoproteins. On the other hand it is just as conceivable, that within specific estrogen target cells alpha-fetoproteins may lose their immunoreactivity, because they may be transformed into some other molecular entity. That such transformation of estrogen-binding proteins is not only possible, but may, in the presence of estradiol, actually alter the binding affinity and generate the active form of an estrogen-receptor, has been demonstrated previously[65]. The fact that alpha-fetoproteins in the serum of immature rats exist in several different molecular forms with different affinities and binding capacities for estradiol[108] proves the ability of alpha-fetoproteins to undergo structural and functional transformations.

The decrease in serum concentrations of alpha-fetoproteins during postnatal development is accompanied by a parallel increase in the concentration of specific intracellular estrogen receptors in the same brain regions, in which immunologically active alpha-fetoproteins could not be detected. This leaves space for the speculation that alpha-fetoproteins, after undergoing multiple conformative changes in the blood circulation, enter estrogen target cells in the brain, where they are transformed into estrogen receptors of even higher affinity. The serum fraction of alpha-fetoprotein with the lowest glycosylation was shown to carry the highest-affinity estrogen binding sites[108]. Taken together with the well-known facilitated entry of deglycosylated glycoproteins into cells, this observation advocates the facilitated entry of those alpha-fetoprotein molecules with the highest-affinity estrogen binding sites into estrogen target cells. The second hypothetical step, intracellular transformation of alpha-fetopro-

116

teins into specific cytoplasmic estrogen receptors, has been shown repeatedly to be reversed at high salt concentration[8, 96, 106].

In summary, estrogens, via their ability to alter the affinity of estrogen-binding sites, may act as catalysts for passing binding proteins through membranes, be it cell membranes or the membranes of the cell nucleus.

Defeminization is linked to impaired function of the estrogen receptor system. It was previously mentioned that female rats, which had been treated pre- and/or postnatally with estrogen antagonists, like tamoxifen or LY 117018, remain permanently irresponsive to later estrogen treatment with regard to the expression of reproductive functions. It was also shown that uptake of ³H-estradiol into the nuclei of preoptic and ventromedial hypothalamic cells was permanently inhibited in these animals[33]. Since uptake and retention of estradiol is predominantly receptor-mediated, these results suggest that development of the estrogen receptor system was permanently impaired. In the adult organism, estrogenic interaction with the genetic material in the cell nucleus stimulates DNA synthesis and stimulates estrogen receptor resynthesis in the cytosol[58, 59, 75]. Since, during the perinatal phase of development, many steroidal effects are of permanent (imprinting) nature, the well-known effect of estrogens on resynthesis of their own receptors in adulthood may have an equivalent of a permanent imprinting nature during development. Since tamoxifen inhibits the stimulatory effects of estrogens on DNA synthesis and estrogen receptor resynthesis in adulthood[58, 75], this estrogen antagonist may act similarly in the developing organism. The result is a permanent impairment in the formation of estrogen receptor systems. It remains uncertain, however, whether the impairment developed because estrogenic interaction with the genetic material was prevented postnatally, or whether tamoxifen may have induced its own receptor system at the expense of the estrogen receptor system. The existence of anti-estrogen binding sites in various estrogen target tissues, which bind tamoxifen, but not estradiol, was recently revealed by Sudo et al.[101].

Clark and Peck[17] previously concluded that the estrogen receptor system develops without the influence of postnatal estrogens. Their conclusion was based on the assumption that gonadectomy would clear the postnatal rat of estrogens. It was, however, discussed previously that postnatal gonadectomy does not clear the developing rat of estrogens (see also ref. 31).

Ontogeny of androgen receptors. Concurrently with the developing estrogen receptor system there is also the development of an androgen receptor system in the hypothalamus[61, 62]. The similarity of estrogen and androgen receptors in the affinity for their respective hormones[20, 72], and the fact that androgen receptors possess affinity for estradiol[43, 72], and estrogen receptors possess affinity for androgens[43, 91, 92], may indicate that both receptor systems may derive from a common undifferentiated precursor. Already Sheridan et al.[93, 94] and Fox[43] have considered the possibility that androgens and estrogens may compete with each other for their receptor during the period of sexual differentiation. Döhler and Hancke[30] demonstrated that hypothalamic masculinization of female rats due to postnatal treatment with testosterone was prevented by concomitant postnatal treatment with estradiol, and vom Saal et al.[111] reported that prenatal estrogens interfered with the effect of androgens in the differentiation of aggressive behavior in male mice. The observation by Csaba (for reviews see ref. 18 and 19) that hormones may induce their own receptors in unicellular organisms may also be valid for receptor induction in higher organisms. There are, however, no studies yet which would prove or disprove the possibility that initial contact of a steroid with a binding protein would imprint this protein in such a way that binding affinity to the particular steroid in question is increased and a specific receptor is formed.

Conclusion

Sexual differentiation of the mammalian brain seems to be more complex than has generally been taken for granted. The available data indicate that the capacity for the display of female sexual behavior and for the cyclic release of gonadotropic hormones is not, as has been assumed, inherent to central nervous tissue. Instead the embryonic brain seems to be as yet undetermined for either a masculine or feminine course of development.

Female differentiation of the brain. Under the influence of moderate levels of estrogens, the embryonic sex centers in the brain differentiate into neural substrates and circuitries which will be able, in adulthood, to respond to female sex hormones (i.e. estrogens and gestagens) with a display of characteristic female sexual functions. During the period of sexual differentiation of the brain moderate levels of estrogens in the fetal blood circulation are provided by the placenta. In rodents with a short gestation period, such as the rat, mouse, and hamster, prenatal estrogens of maternal or placental origin are carried over into the postnatal period, probably by alpha-fetoproteins, which protect the estrogens from metabolic degradation. Although the majority of these estrogens circulate in a biologically inactive form, they are always 'on-call' and are immediately available at the target site either by rapid dissociation from the estrogen-alpha-fetoprotein complex, or by catalyzing the entry of alpha-fetoprotein into target cells.

Male differentiation of the brain. Under the influence of sufficiently high levels of estrogens, on the other hand, the embryonic sex centers in the brain will develop and differentiate into neural substrates and circuitries which are able, in adulthood, to respond physiologically in a male fashion to estrogens and to aromatizable androgens. High levels of estrogens are delivered perinatally for local action on the brain by means of aromatizable androgens, released from the developing testes. Since androgens are not bound by alpha-fetoproteins, they can reach the brain fairly rapidly, and be aromatized intracellularly into estrogens within certain brain areas. Via this mechanism, estrogens can act specifically and at high concentrations on selective brain regions without influencing other estrogen-sensitive target tissues. Structural differentiation of the male brain seems to depend exclusively on the priming action of estrogens.

For masculinization and defeminization of brain functions, however, the available data suggest that androgens per se may be more important, and estrogens

118

per se may be less important, than has been assumed. Additional information about molecular mechanisms beyond steroid-induced transcription, and about the possible influence of neurotransmitters and neuropeptides, is needed for further elucidation of the mechanism of hormone induced sexual differentiation of the brain.

Sexual differentiation: controled by receptor imprinting? It is reasonable to assume that steroid receptors are involved in the process of steroid-induced sexual differentiation. In prenatal rats specific estrogen receptors are still undetectable, but they increase in number substantially in hypothalamus, amygdala and cortex during the postnatal period. Since, during the perinatal phase of development, many steroidal effects are of permanent (imprinting) nature, the well-known effect of estrogens on resynthesis of their own receptors in adulthood may have an equivalent of a permanent imprinting nature during development. Such an imprinting mechanism could explain the ontogeny of different affinity forms of estrogen-binding proteins in serum and in estrogen target cells and non-target cells against a background of greatly elevated circulating estrogen levels. The serum fraction of alpha-fetoprotein with lowest glycosylation was shown to carry the highest-affinity estrogen binding sites. Since deglycosylated glycoproteins show facilitated entry into cells, it is very suggestive that the alpha-fetoprotein molecules with the highest affinity estrogen binding sites will preferentially enter estrogen target cells. In other words, the binding affinity for estradiol increases the closer the binding protein gets to the target cell and to the nucleus respectively. Thus, the perinatal rat possesses an estrogen binding system, which prevents degradation of estrogens in peripheral organs, but conveys estrogens through its affinity gradient directly into target cells. Another possibility is that estrogens, via their ability to alter the affinity of estrogen-binding sites, may act as catalysts for passing binding proteins through membranes and, thus, may convey alpha-fetoproteins into estrogen traget cells. It should be mentioned, however, that there are no studies yet which would prove or disprove the possibility that initial contact of a steroid with a binding protein would imprint this protein in such a way that binding affinity to the particular steroid in question is increased and a specific receptor is formed.

Acknowledgments. I am very grateful to the many colleagues and friends in various countries who engaged with me in fruitful discussions on this topic, and thus stimulated the generation of new ideas. I am especially grateful to my co-workers J. L. Hancke (Valdivia/Chile), B. Jarzab (Zabrze/Poland), A. Sipos (Budapest/Hungary), S. S. Srivastava (Jaunpur/India), and C. C. Wong (Hong Kong), A. Coquelin, F. Davis, R. A. Gorski, M. Hines, and J. E. Shryne (Los Angeles/USA), E. Nordeen, and P. Yahr (Irvine/USA), C. Ganzemüller, C. Hofmann, P. M. Sickmöller, and C. Veit (Hannover/FRG), and to M. König (Bissendorf/FRG), who prepared and organized all the necessary paper work for this review.

1   Anaxagoras, quoted by Aristotle. Generation of Animals, Book IV, Chap. I, 736b, pp. 32–36. Translated by A. L. Peck, Loeb Classical Library, Heinemann, London 1953.
2   Anderson, J. N., Peck, E. J., and Clark, J. H., Nuclear receptor estrogen complex: accumulation, retention and localization in the hypothalamus and pituitary. Endocrinology 93 (1973) 711–717.
3   Arai, Y., and Gorski, R. A., Critical exposure time for androgenization of the rat hypothalamus determined by anti-androgen injection. Proc. Soc. expl Biol. Med. 127 (1968) 590–593.
4   Aristotle. History of Animals, Book VI, Chap. XIX/2, p. 165. Translated by R. Cresswell, H. G. Bohn, London 1862.

5   Attardi, B., Geller, L. N., and Ohno, S., Androgen and estrogen receptor in brain cytosol from male, female and testicular feminized (tfm) mice. Endocrinology *98* (1976) 864–874.

6   Attardi, B., and Ruoslahti, E., Foetoneonatal estradiol-binding protein in mouse brain cytosol is α–foetoprotein. Nature *263* (1976) 685–687.

7   Aussel, C., Uriel, J., Michel, G., and Baulieu, E. E., Immunological demonstration of α-fetoprotein in uterine cytosol from immature rats. Biochimie *56* (1974) 567–570.

8   Bayard, B., Kerckaert, J. P., and Biserte, G., Differences in the molecular heterogeneity of alpha-fetoprotein from uterus and serum of immature rats. Biochem. biophys. Res. Commun. *85* (1978) 47–54.

9   Belisle, S., and Tulchinsky, D., Amnionic fluid hormones, in: Maternal-fetal Endocrinology, pp. 169–195. Eds D. Tulchinsky and K. Ryan. Saunders. Philadelphia, PA 1980.

10   Benno, R., and Williams, T., Evidence for intracellular localization of alpha-fetoprotein in the developing rat brain. Brain Res. *142* (1978) 182–186.

11   Booth, J. E., Sexual behaviour of male rats injected with the antioestrogen MER-25 during infancy. Physiol. Behav. *19* (1977) 35–39.

12   Booth, J. E., Sexual behavior of neonatally castrated rats injected during infancy with oestrogen and dihydrotestosterone. J. Endocr. *72* (1977) 135–141.

13   Booth, J. E., Sexual differentiation of the brain, in: Oxford Reviews of Reproductive Biology, vol. 1, pp. 58–158. Ed. C. A. Finn, Clarendon Press, Oxford 1979.

14   Chang, C. Y., and Witschi, E., Independence of adrenal hyperplasia and gonadal masculinization in the experimental adrenogenital syndrome of frogs. Endocrinology *56* (1955) 597–605.

15   Chang, C. Y., and Witschi, E., Genetic control and hormonal reversal of sex differentiation in Xenopus. Proc. Soc. expl. Biol. Med. *93* (1956) 140–144.

16   Clark, J. H., and McCormack, S., Clomid or nafoxidine administered to neonatal rats causes reproductive tract abnormalities. Science *197* (1977) 164–165.

17   Clark, J. H., and Peck, E. J. Jr, Female sex steroids: receptors and function, Monographs on Endocrinology, Springer Verlag, Berlin 1979.

18   Csaba, G., Ontogeny and Phylogeny of Hormone Receptors, Karger, Basel 1981.

19   Csaba, G., Receptor ontogeny and hormonal imprinting. Experientia *42* (1986) 750–759.

20   Davies, J., Siu, J., Naftolin, F., and Ryan, K. J., Cytoplasmic binding of steroids in brain tissues and pituitary, in: Advances in the Biosciences, 15th Schering Workshop on Central Actions of Estrogenic Hormones, pp. 89–103. Ed. G. Raspé. Pergamon Press, Vieweg 1975.

21   De Vries, G. J., Buijs, R. M., and Swaab, D. F., Ontogeny of the vasopressinergic neurons of the suprachiasmatic nucleus and their extrahypothalamic projection in the rat brain – presence of a sex difference in the lateral septum. Brain Res. *218* (1981) 67–78.

22   De Vries, G. J., Buijs, R. M., and Van Leeuwen, F. W., Sex differences in vasopressin and other neurotransmitter systems in the brain. Progr. Brain Res. *61* (1984) 185–203.

23   Döhler, K.-D., Is female sexual differentiation hormone mediated? Trends Neurosci. *1* (1978) 138–140.

24   Döhler, K.-D., Coquelin, A., Davis, F., Hines, M., Shryne, J. E., and Gorski, R. A., Differentiation of the sexually dimorphic nucleus in the preoptic area of the rat brain is determined by the perinatal hormone environment. Neurosci. Lett. *33* (1982) 295–298.

25   Döhler, K.-D., Coquelin, A., Davis, F., Hines, M., Shryne, J. E., and Gorski, R. A., Geschlechtsunterschiede in der Grobstruktur des Rattenhirns und ihre Prägung durch Sexualhormone. Tierärztl. Prax. *11* (1983) 543–550.

26   Döhler, K.-D., Coquelin, A., Davis, F., Hines, M., Shryne, J. E., and Gorski, R. A., Pre- and postnatal influence of testosterone propionate and diethylstilbestrol on differentiation of the sexually dimorphic nucleus of the preoptic area in male and female rats. Brain Res. *302* (1984) 291–295.

27   Döhler, K.-D., Coquelin, A., Davis, F., Hines, M., Shryne, J. E., Sickmöller, P. M., Jarzab, B., and Gorski, R. A., Pre- and postnatal influence of an estrogen antagonist and an androgen antagonist on differentiation of the sexually dimorphic nucleus of the preoptic area in male and female rats. Neuroendocrinology, *42* (1986) 443–448.

28   Döhler, K.-D., Coquelin, A., Hines, M., Davis, F., Shryne, J. E., and Gorski, R. A., Hormonal influence on sexual differentiation of rat brain anatomy, in: Hormones and Behavior in Higher Vertebrates, pp. 194–203. Eds J. Balthazart, E. Pröve and R. Gilles. Springer, Berlin 1983.

29   Döhler, K.-D., and Hancke, J. L., Thoughts on the mechanism of sexual brain differentiation, in: Hormones and Brain Development, pp. 153–158. Eds G. Dörner and M. Kawakami. Elsevier/North-Holland Biomedical Press, Amsterdam 1978.

30 Döhler, K.-D., and Hancke, J. L., Testosterone-induced hypothalamic masculinization of female rats is prevented by estradiol and is augmented by dihydrotestosterone. Acta endocr., suppl. *225* (1979) 245.

31 Döhler, K.-D., Hancke, J. L., Srivastava, S. S., Hofmann, C., Shryne, J. E., Gorski, R. A., Participation of estrogens in female sexual differentiation of the brain; neuroanatomical neuroendocrine and behavioral evidence. Progr. Brain Res. *61* (1984) 99–117.

32 Döhler, K.-D., Hines, M., Coquelin, A., Davis, F., Shryne, J. E., and Gorski, R. A., Pre- and postnatal influence of diethylstilboestrol on differentiation of the sexually dimorphic nucleus in the preoptic area of the female rat brain. Neuroendocr. Lett. *4* (1982) 361–365.

33 Döhler, K.-D., Nordeen, E. J., and Yahr, P., The uptake of ^{3}H-estradiol into brain cell nuclei is permanently inhibited in rats after perinatal treatment with tamoxifen. Acta endocr. *102* suppl. 253 (1983) 47–48.

34 Döhler, K.-D., Srivastava, S. S., Shryne, J. E., Jarzab, B., Sipos, A., and Gorski, R. A., Differentiation of the sexually dimorphic nucleus in the preoptic area of the rat brain is inhibited by postnatal treatment with an estrogen antagonist. Neuroendocrinology *38* (1984) 297–301.

35 Döhler, K.-D., von zur Mühlen, A., Döhler, U., and Fricke, E., Development of the pituitary-thyroid axis in male and female rats. Acta endocr. *84*, suppl. 208 (1977) 2-3.

36 Döhler, K.-D., and Wuttke, W., Changes with age in levels of serum gonadotropins, prolactin, and gonadal steroids in prepubertal male and female rats. Endocrinology *97* (1975) 898–907.

37 Dörner, G., Sexual differentiation of the brain. Vitam. Horm. *38* (1981) 325–381.

38 Dörner, G., and Staudt, J., Structural changes in the preoptic anterior hypothalamic area of the male rat, following neonatal castration and androgen substitution. Neuroendocrinology *3* (1968) 136–140.

39 Dörner, G., and Staudt, J., Structural changes in the hypothalamic ventromedial nucleus of the male rat, following neonatal castration and androgen treatment. Neuroendocrinology *4* (1969) 278–281.

40 Dyer, R. G., MacLeod, N. K., and Ellendorf, F., Electrophysiological evidence for sexual dimorphism and synaptic convergence in the preoptic and anterior hypothalamic areas of the rat. Proc. R. Soc. *193* (1976) 421–440.

41 Dyer, R. G., Sexual differentiation of the forebrain – relationship to gonadotrophin secretion. Progr. Brain Res. *61* (1984) 223–236.

42 Everett, J. W., Sawyer, C. H., and Markee, J. E., A neurogenic timing factor in control of the ovulatory discharge of luteinizing hormone in the cycling rat. Endocrinology *44* (1949) 234–250.

43 Fox, T. O., Androgen- and estrogen-binding macromolecules in developing mouse brain: biochemical and genetic evidence. Proc. natn. Acad. Sci. USA *72* (1975) 4303–4307.

44 Gellert, R. J., Bakke, J. L., and Lawrence, N. L., Persistent estrus and altered estrogen sensitivity in rats treated neonatally with clomiphene citrate. Fert. Steril. *22* (1971) 244–250.

45 Gibori, G., and Sidaran, R., Sites of androgen and estradiol production in the second half of pregnancy in the rat. Biol. Reprod. *24* (1981) 249–256.

46 Gladue, B. A., and Clemens, L. G., Androgenic influences on feminine sexual behavior in male and female rats: defeminization blocked by prenatal antiandrogen treatment. Endocrinology *103* (1978) 1702–1709.

47 Gorski, R. A., Critical role of the medial preoptic area in the sexual differentiation of the brain. Progr. Brain Res. *61* (1984) 129–146.

48 Gorski, R. A., Csernus, V. J., and Jacobson, C. D., Sexual dimorphism in the preoptic area, in: Advances in Physiological Sciences; Reproduction and Development, vol. 15, pp. 121–130. Eds B. Flerkó, G. Sétáló and L. Tima. Pergamon Press and Akadémia Kiadó Press, Budapest 1980.

49 Gorski, R. A., Gordon, J. H., Shryne, J. E., and Southam, A. M., Evidence for a morphological sex difference within the medial preoptic area of the rat brain. Brain Res. *148* (1978) 333–346.

50 Gorski, R. A., and Jacobson, C. D., Sexual differentiation of the brain, in: Pediatric Andrology, pp. 109–134. Eds S. J. Kogan and E. S. E. Hafez. Martinus Nijhoff, The Hague 1981.

51 Goy, R. W., and McEwen, B. S., Sexual Differentiation of the Brain. The MIT Press, Cambridge, Mass. 1980.

52 Greenough, W. T., Carter, C. S., Steerman, C., and De Voogt, T. J., Sex differences in dendritic patterns in hamster preoptic area. Brain Res. *126* (1977) 63–72.

53 Hancke, J. L., and Döhler, K.-D., Sexual differentiation of female brain function is prevented by postnatal treatment of rats with the estrogen antagonist tamoxifen. Neuroendocr. Lett. *6* (1984) 201–206.

54 Hsü, H. K., Chen, F. N., and Peng, M. T., Some characteristics of the darkly stained area of the medial preoptic area of rats. Neuroendocrinology *31* (1980) 327–330.

55 Jacobson, C. D., Csernus, V. J., Shryne, J. E., and Gorski, R. A., The influence of gonadectomy, androgen exposure, or a gonadal graft in the neonatal rat on the volume of the sexually dimorphic nucleus of the preoptic area. J. Neurosci. *1* (1981) 1142–1147.

56 Jacobson, C. D., and Gorski, R. A., Neurogenesis of the sexually dimorphic nucleus of the preoptic area in the rat. J. comp. Neurol. *196* (1981) 519–529.

57 Jacobson, C. D., Shryne, J. E., Shapiro, F., and Gorski, R. A., Ontogeny of the sexually dimorphic nucleus of the preoptic area. J. comp. Neurol. *193* (1980) 541–548.

58 Jordan, V. C., Dix, C. J., Rowsby, L., and Prestwich, G., Studies on the mechanism of action of the nonsteroidal antiestrogen tamoxifen in the rat. Molec. cell. Endocr. *7* (1977) 177–192.

59 Jordan, V. C., Prestwich, G., Dix, C. J., and Clark, E. R., Binding of antie-strogens to the estrogen receptor, the first step in anti-estrogen action, in: Pharmacological Modulation of Steroid Action, pp. 81–98. Eds E. Genazzani, F. DiCarlo and W. I. P. Mainwaring. Raven Press, New York 1980.

60 Jost, A., Sur le controle hormonal de differenciation sexuelle du lapin. Archs Anat. micr. Morph. expl *39* (1950) 577–598.

61 Kato, J., Cytosol and nuclear receptors for 5α-dihydrotestosterone and testosterone in the hypothalamus and hypophysis, and testosterone receptors from neonatal female rat hypothalamus. J. steroid Biochem. *7* (1976) 1179–1187.

62 Kato, J., Ontogeny of 5α-dihydrotestosterone receptors in the hypothalamus of the rat. Ann. Biol. anim. Biochim. Biophys. *16* (1976) 467–469.

63 Kime, D., Vinson, G., Major, P., and Kilpatrick, R., Adrenal-gonadal relationships, in: General, Comparative and Clinical Endocrinology of the Adrenal Cortex, vol. 3, pp. 183–264. Eds I. Jones and I. Henderson. Academic Press, New York 1980.

64 Lamprecht, S. A., Kohen, F., Ausher, J., Zor, U., and Lindner, H., Hormonal stimulation of estradiol-17β release from the rat ovary during early postnatal development. J. Endocr. *68* (1976) 343–344.

65 Little, M., Szendro, C., Teran, C., Hughes, A., and Jungblut, P. W., Biosynthesis and transformation of microsomal and cytosol estradiol receptors. J. steroid Biochem. *6* (1975) 493–500.

66 MacLusky, N. J., Lieberburg, I., and McEwen, B. S., Development of steroid receptor systems in the rodent brain, in: Ontogeny of Receptors and Reproductive Hormone action, pp. 393–402. Eds T. H. Hamilton, J. H. Clark and W. A. Sadler. Raven Press, New York 1979.

67 Mainwaring, W. I. P., Modes of action of antiandrogens: a survey, in: Androgens and Antiandrogens, pp. 151–161. Eds L. Martini and M. Motta. Raven Press, New York 1977.

68 McEwen, B. S., Chaptal, C., Gerlach, J., and Wallach, G., The role of fetoneonatal estrogen binding proteins in the associations of estrogen with neonatal brain cell nuclear receptors. Brain Res. *96* (1975) 400–406.

69 McEwen, B. S., Lieberburg, I., Chaptal, C., and Krey, L. C., Aromatization: important for sexual differentiation of the neonatal rat brain. Horm. Behav. *9* (1977) 249–263.

70 Migeon, C. J., Amrhein, J. A., Keenan, B. S., Meyer, W. J. III, and Migeon, B. R., The syndrome of androgen insensitivity in man: its relation to our understanding of male sex differentiation, in: Genetic Mechanisms of Sexual Development, pp. 93–147. Eds H. L. Vallet and I. H. Porter. Academic Press, New York 1979.

71 Migeon, C. J., Brown, T. R., and Fichman, K. R., Androgen insensitivity syndrome, in: The Intersex Child; Pediatric and Adolescent Endocrinology, vol. 8, pp. 171–202. Ed. N. Josso. Karger, Basel 1981.

72 Naess, O., Hansson, V., Djoeseland, O., and Attramadal, A., Characterization of the androgen receptor in the anterior pituitary of the rat. Endocrinology *97* (1975) 1355–1363.

73 Neumann, F., and Elger, W., Permanent changes in gonadal function and sexual behavior as a result of early feminization of male rats by treatment with an antiandrogenic steroid. Endokrinologie *50* (1966) 209–224.

74 Neumann, F., and Kramer, M., Female brain differentiation of male rats as a result of early treatment with an androgen antagonist, in: Hormonal Steroids, pp. 932–941. Eds L. Martini, F. Fraschini and M. Motta. Excerpta Medica, Amsterdam 1967.

75 Nicholson, R. I., Golder, M. P., Davies, P., and Griffiths, K., Effects of oestradiol-17β and tamoxifen on total and accessible cytoplasmatic oestradiol-17β-receptors in DMBA-induced rat mammary tumours. Eur. J. Cancer *12* (1976) 711–717.

76 Nishizuka, M., and Arai, Y., Sexual dimorphism in synaptic organization in the amygdala and its dependence on neonatal hormone environment. Brain Res. *212* (1981) 31-38.

77 Nottebohm, F., and Arnold, A. P., Sexual dimorphism in vocal control areas of the songbird brain. Science *194* (1976) 211–213.

78   Nunez, E. A., Engelmann, F., Benassayag, C., Savu, L., Crepy, O., and Jayle, M. F., Mise en évidence d'une fraction protéique liant les oestrogènes dans le sérum de rats impuberes. C.r. Acad. Sci. Paris D *272* (1971) 2396–2399.

79   Peterson, R. E., Imperato-McGinley, J., Gautier, T., and Sturla, E., Hereditary steroid 5α-reductase deficiency: a newly recognized cause of male pseudohermaphroditism, in: Genetic Mechanisms of Sexual Development, pp. 149–167. Eds H. L. Vallet and I. H. Porter. Academic Press, New York 1979.

80   Pfaff, D. W., Morphological changes in the brains of adult male rats after neonatal castration. J. Endocr. *36* (1966) 415–416.

81   Pfeiffer, C. A., Sexual differences of the hypophyses and their determination by the gonads. Am. J. Anat. *58* (1936) 195–226.

82   Pieau, C., Temperature and sex differentiation in embryos of two chalonians, *Emys orbicularis L.* and *Testudo graeca L.,* in: Intersexuality in the Animal Kingdom, pp. 332–339. Ed. R. Reinboth. Springer, New York 1975.

83   Piquet, J., Détermination de sexe chez les batraciens en fonction de la température. Rev. Suisse Zool. *37* (1930) 173–281.

84   Plapinger, L., and McEwen, B. S., Gonadal steroid-brain interactions in sexual differentiation, in: Biological Determinants of Sexual Behaviour, pp. 153–218. Ed. J. B. Hutchison. John Wiley & Sons, New York 1978.

85   Plato. Symposium, in: The dialogues of Plato, vol. 1, 4th edn, 189d–190a, p. 521. Translated by B. Jowett. Clarendon Press, Oxford 1953.

86   Plato. Timaeus, 90E–91A, p. 248, Loeb Classical Library London, Heinemann, London 1975.

87   Raisman, G., and Fields, P. M., Sexual dimorphism in neuropil of the preoptic area of the rat and its dependence on neonatal androgen. Brain Res. *54* (1973) 1–29.

88   Raynaud, J. P., Mercier-Bodard, C., and Beaulieu, E. E., Rat estradiol binding plasma protein. Steroids *18* (1971) 767–787.

89   Resko, J. A., Ploem, J. G., and Stadelman, H. L., Estrogens in fetal and maternal plasma of the rhesus monkey. Endocrinology *97* (1975) 425–430.

90   Reyes, F. I., Boroditsky, R. S., Winter, J. S. D., and Faiman, C., Studies on human sexual development. II. Fetal and maternal serum gonadotropin and sex steroid concentrations. J. clin. Endocr. Metab. *38* (1974) 612–617.

91   Rochefort, H. L., Lignon, F., and Capony, F., Formation of estrogen nuclear receptors in uterus: effect of androgens, estrone and nafoxidine. Biochem. biophys. Res. Commun. *47* (1972) 662.

92   Schmidt, W., Sandler, M. A., and Katzenellenbogen, B. W., Androgen-uterine interaction: nuclear translocation of the estrogen receptor and induction of the synthesis of the uterine-induced protein (IP) by high concentrations of androgens *in vitro* but not *in vivo*. Endocrinology *98* (1976) 702–716.

93   Sheridan, P. J., Sar, M., and Stumpf, W. E., Autoradiographic localization of 3H-estradiol or its metabolites in the central nervous system of the developing rat. Endocrinology *94* (1974) 1386–1390.

94   Sheridan, P. J., Sar, M., and Stumpf, W. E., Interaction of exogenous steroids in the developing rat brain. Endocrinology *95* (1974) 1749–1753.

95   Short, R. V., Sex determination and differentiation, in: Reproduction in Mammals, vol. 2, pp. 70–113. Eds C. R. Austin and R. V. Short. Cambridge University Press, Cambridge 1982.

96   Smalley, J. R., and Sarcione, E. J., Synthesis of alpha-fetoprotein by immature rat uterus. Biochem. biophys. Res. Commun. *92* (1980) 1429–1434.

97   Södersten, P., Effects of anti-oestrogen treatment of neonatal male rats on lordosis behaviour and mounting behaviour in the adult. J. Endocr. *76* (1978) 241–249.

98   Stolkovski, J., and Bellec, A., Influence du rapport potassium/calcium du milieu d'élevage sur la distribution des sexes chez *Discoglosus pictus* (Otth). C.r. Acad. Sci. Paris *251* (1960) 1669–1671.

99   Strott, C. A., Sundrel, H., and Stahlman, M. L., Maternal and fetal plasma progesterone, cortisol, testosterone and 17β-estradiol in preparaturient sheep: response to fetal ACTH infusion. Endocrinology *95* (1974) 1327–1339.

100   Stumpf, W. E., Sar, M., and Keefer, D. A., Atlas of estrogens target cells-rat brain, in: Anatomical Neuroendocrinology, pp. 104–119. Eds W. E. Stumpf and L. D. Grant. Karger, Basel 1975.

101   Sudo, K., Monsma, F. J. Jr, and Katzenellenbogen, B. S., Antiestrogenbinding sites distinct from the estrogen receptor: subcellular localization, ligand specificity, and distribution in tissues of the rat. Endocrinology *112* (1983) 425–434.

102 Szalay, R., Krieg, M., Schmidt, H., and Voigt, K.-D., Metabolism and mode of action of androgens in target tissues of male rats. Acta endocr. Copenh., suppl. *80* (1975) 592–602.

103 Toran-Allerand, C. D., Sex steroids and the development of the newborn mouse hypothalamus and preoptic area in vitro: implications for sexual differentiation. Brain Res. *106* (1976) 407–412.

104 Toran-Allerand, C. D., Regional differences in intraneuronal localization of alpha-fetoprotein in developing mouse brain. Devl. Brain Res. *5* (1982) 213–217.

105 Toran-Allerand, C. D., On the genesis of sexual differentiation of the central nervous system: morphogenetic consequence of steroidal exposure and possible role of alpha-fetoprotein. Progr. Brain Res. *61* (1984) 63–98.

106 Uriel, J., Tissular receptors of estrogens and AFP, in: Carcino-embryonic Proteins, vol. I, pp. 181–189. Ed. F.-G. Lehmann. Elsevier/North-Holland Biomedical Press, Amsterdam 1979.

107 Uriel, J., Bouillon, D. Aussel, C., and Dupier, M., Alpha-fetoprotein: the major high-affinity estrogen binder in rat uterine cytosols. Proc. natn. Acad. Sci. USA *73* (1976) 1452–1456.

108 Vallette, G., Benassayag, C., Belanger, L., Nunez, E. A., and Jayle, M. F., Rat iso-alpha-fetoproteins: purification and interaction with estradiol-17β. Steroids *29* (1977) 277–289.

109 Veit, C., Ganzemüller, C., and Döhler, K. D., Postnatal treatment of male rats with the estrogen antagonist LY 117018 increases the capacity for lordosis behavior. First Congr. comp. Physiol. Biochem., Liège, abstract D26.t, 1984.

110 Vom Saal, F. S., The interaction of circulating estrogens and androgens in regulating mammalian sexual differentiation, in: Hormones and Behaviour in Higher Vertebrates, pp. 194–203. Eds J. Balthazart, E. Pröve and R. Gilles. Springer, Berlin 1983.

111 Vom Saal, F. S., Grant, W. M., McMullen, C. W., and Laves, K. S., High fetal estrogen concentrations: correlation with increased adult sexual performance and decreased agression in male mice. Science *220* (1983) 1306–1309.

112 Vreeburg, J. T. M., van der Vaart, P. D. M., and van der Schoot, P., Prevention of central defeminization, but not masculinization in male rats by inhibition neonatally of oestrogen biosynthesis. J. Endocr. *74* (1977) 375–382.

113 Weisz, J., and Gunsalus, P., Estrogen levels in immature female rats: true or spurious – ovarian or adrenal? Endocrinology *93* (1973) 1057–1065.

114 Yamamoto, T., Artificial induction of functional sex reversal in genotypic females of the medaka *(Oryzias latipes)*. J. expl Zool. *137* (1958) 227–263.

115 Yamamoto, T., A further study of induction of functional sex reversal in genotypic males of the medaka *(Oryzias latipes)* and progenies of sex reversals. Genetics *44* (1959) 739–757.

116 Yamamoto, T., Progenies of induced sex-reversal females mated with induced sex-reversal males in the medaka *(Oryzias latipes)*. J. expl Zool. *146* (1963) 163–180.

The mechanism of receptor development as implied by hormonal imprinting studies on unicellular organisms

P. Kovács

Department of Biology, Semmelweis University of Medicine, POB 370, H–1445 Budapest (Hungary)

In higher organisms, the development of hormone receptors is genetically encoded and forms an integral part of cell membrane differentiation[6]. Full maturation of the receptor nevertheless requires the presence of the hormone for amplification (hormonal imprinting)[4,7]. Hormonal imprinting takes place at the primary interaction between target cell and hormone, and amplifies the cellular interacting structure (receptor) for specific binding of the hormone. It has been shown experimentally that hormonal imprinting can also occur in unicellular organisms[4,6] if the hormone is present in their environment[3]. Thus the phenomenon of hormonal imprinting can be reproduced experimentally in unicellular model systems[7].

The unicellular organism *Tetrahymena* has proved to be an ideal model organism for hormone receptor studies from several points of view. First and foremost, the *Tetrahymena* cells do not require separation for culturing, which ensures the integrity of their membrane, and they can be easily maintained and propagated in culture. A still greater advantage is that preformed specific receptors are relatively numerous in the *Tetrahymena,* and the inducibility and formation of binding sites (receptors) for hormones can therefore be relatively easily studied by adequate methods. Evidence has also been obtained that the membrane receptors of the *Tetrahymena* are in many respects similar to those of higher organisms, e.g. of mammals. A cautious extrapolation of observation on the receptors of the *Tetrahymena* to those of higher organisms is therefore possible.

The *Tetrahymena* is a free-living ciliated unicellular organism for which adaptation to environmental changes is vitally important. Adaptation presupposes the presence in the cell membrane of receiver structures capable of recognizing environmental changes, and of mediating these to an effector system which brings about an appropriate cellular response(s)[6]. Such receiver structures may play a role in the uptake of nutrients, recognition of poisonous materials (molecules), and of individuals of the opposite sex in the case of sexually reproducing species, etc. Since the environmental changes are very varied, and in some cases also abrupt, the cellular responses are necessarily similarly prompt and manifold[5]. The environmental changes presumably alter the membrane itself, in a manner that enables it to respond to the next similar change more intensively and promptly than it originally responded. We believe that such events are involved in hormone receptor formation, too. The environmental changes act not only on the cell membrane, but also on the entire system of cellular functions, activating thereby an effector mechanism which controls, at transcription level, the intracellular response to the stimulus received. For example, the presence of metals in the environment of *Tetrahymena* causes it to synthesize metal-binding proteins[44].

The above implications have been substantiated by experimental observations[2]. We have been able to induce in the membrane of the *Tetrahymena* specific-appearing receptors to certain materials (hormones), which neither are, nor have ever been present in its natural environment. The experimental induction of structures acting as receptors permits an insight into the mechanism of receptor formation.

Although several hormones and hormone-like materials have been demonstrated in the body of *Tetrahymena*[41, 42], it seems unlikely that this protozoon possesses preformed receptors for vertebrate hormones. According to present knowledge, few receptor structures can be regarded as preformed at the unicellular level. A preformed structure is, for example, the nutrient receptor[40], which accounts primarily for peptide binding; the polypeptide-type hormone receptors presumably developed from the primitive nutrient receptor structures. However, certain experimental observations suggest that receptors for hormones other than those of a polypeptide type can also occur in the membrane of *Tetrahymena*[24, 25, 28, 31].

The substantiation of hormonal imprinting, and detection of the specific hormone binding sites, requires extremely sensitive procedures, and the study of adequate parameters which can show conclusively an increase (or a decrease) in the number of receptors. Such procedures are on the one hand methods for the detection of hormone binding, and on the other, those assessing certain physiological responses which portray the quantitative relations of hormone reception (phagocytic activity, growth rate, metabolic rate, ect.). Much valuable information has also been derived from investigations into changes in the quantitative relations of hormone receptor components, e.g. lectin binding studies can precisely reveal quantitative alterations of the saccharide components of membrane receptors[10].

The supposition that the hormone(s) added to the maintenance medium of the *Tetrahymena* cells act by mediation of the cell membrane leads to the presumption that the hormone-induced changes depend on the following factors: 1) physiological state, composition, and fluidity of the membrane; 2) integrity of the membrane-associated adenyl-cyclase-cAMP-cPDE and guanylate cyclase systems; 3) quantity and undisturbed operation of other second messengers, such as Ca^{2+}, inositol trisphosphate, etc.; 4) appropriate functioning of the calmodulin system; 5) whether the hormonal imprinting persists over many subsequent generations in which case the collaboration of certain nuclear regions is highly possible. Thus the hormone-induced changes presumably also depend on nuclear functions and on the undisturbed course of protein synthesis, which are indispensable for receptor formation.

The constancy of the culture conditions is an essential prerequisite of investigations into the factors influencing hormone action, because changes in the environmental conditions (ambient temperature, photoperiod, nutrient supply, osmotic conditions) can greatly influence the composition of the cell membrane, and thereby the inducibility of imprinting as well[30, 35].

The role of the cell membrane in hormonal imprinting

Imprinting by polypeptide hormones acting on membrane receptors can take place only in physiological conditions of membrane composition and fluidity. Changes in membrane fluidity and culturing temperature (which alter the relative proportion of unsaturated fatty acids)[53], or in the membrane steroids (e.g. replacement of membrane tetrahymenol by ergosterol, which accounts for rigidity of the membrane) interfere with the normal course of imprinting[37]. Membrane fluidity and, consequently, the mechanism of imprinting, is also affected by local anesthetics and phenothiazine derivatives[45]. Those membrane components which are integral parts of the receiver structures, for example certain saccharides, play a major role in hormonal imprinting. The membrane of the *Tetrahymena* binds concanavalin-A (Con-A), and since the insulin receptor contains saccharide components which are also able to Con-A, pretreatment with Con-A inhibits insulin binding to fixed cells. However, if Con-A and insulin treatment are applied either simultaneously, or on an overlapping scheme, the progeny generations show a significantly greater binding affinity for both ligands than on treatment with either ligand in itself[33]. Similar observations were made in hepatoma cell cultures, in which exposure to Con-A increased not only the binding of insulin to the membrane, but also the internalization of insulin[51]. Con-A presumably enhances the formation of additional binding sites for peptides (insulin), and accounts thereby for a greater intensity of imprinting. It appears that Con-A and insulin act on cPDE-activity, which is one active factor of the Ca^{2+}-calmodulin system, not only in vertebrate target cells, but also in the unicellular *Tetrahymena*[52].

Hormone-membrane interaction seems to be an indispensable prerequisite of imprinting for those hormones which act on membrane receptors. For example, imprinting failed to take place if insulin was administered into the cytoplasm of the *Tetrahymena* in a liposome-coated form, which excluded a membrane-hormone interaction[47].

Agents causing cell membrane perturbation, for example endotoxins, also prevent hormonal imprinting. However, the same agents do not appreciably interfere with hormonal imprinting if they are applied 4 h after hormone exposure. It appears that the establishment of imprinting requires a certain time, and no membrane perturbation can extinguish it once it has become established[46].

Agents known to inhibit clustering during receptor mediated endocytosis, such as methylamine, etc., do not inhibit imprinting[39].

The above experimental observations indicate that the main prerequisite of imprinting is the physiological integrity of the cell membrane, because changes in its normal physical-chemical state inhibit the development of imprinting.

The role of second messengers in hormonal imprinting

The second messengers, such as Ca^{2+}, cAMP, cGMP, etc., play a key role in mediation of the hormonal signal. Since hormonal imprinting presumably involves mediation to the nucleus of the information carried by the signal (hormone) molecule, dysfunction of the intracellular effector system hampers the normal course of hormonal imprinting.

Binding of the hormone to the membrane receptor alters the Ca^{2+}-binding capacity of the membrane[49]. It is known that insulin depresses, whereas glucagon enhances the binding of Ca^{2+}. Inhibition of Ca-binding with La^{3+} does not interfere with the insulin-imprinting of *Tetrahymena* cells (insulin depresses itself the binding of Ca) but prevents the binding of TSH[34]. Conversely, EDTA and EGTA, which form chelates with Ca^{2+} and Mg^{2+}, and alter the influence of the ciliary-membrane-associated guanylate-cyclase-calmodulin complex on the intracellular Ca-level of the *Tetrahymena*[50], inhibit imprinting by insulin, but enhance rather than depress imprinting by TSH, to judge from a significant increase in TSH-binding after imprinting in the presence of EDTA or EGTA[34]. EDTA also accounted for a roughly 100% increased in the hepatocellular response to glucagon[1].

TMB-8, which prevents the establishment of a normal intracellular Ca^{2+}-level, and the Ca^{2+}-antagonizing Ni^{2+} ions as well, have been recognized as further inhibitors of hormonal imprinting[45].

Endogenous cAMP acts as second messenger also in the *Tetrahymena*. Exogenous cAMP modifies the mechanism of imprinting. Treatment of the *Tetrahymena* with dibutyryl-cAMP, or with the cPDE-inhibitor theophylline, equally inhibits imprinting by insulin and TSH[12].

Lithium ions inhibit specifically the action of TSH on the thyroid cell membrane[1], and also inhibit imprinting of the *Tetrahymena* by TSH[12].

The experimental observations outlined above unequivocally suggest that imprinting is inhibited by any gross interference with the function of the intracellular system responsible for mediation of the signals received by the cell membrane. The normal course of imprinting is severely disturbed by any dysfunction of the second messengers (Ca^{2+}; cAMP).

Impact of the inhibition of endocytosis, transcription and translation on hormonal imprinting

The influence of transcription and translation inhibitors on hormonal imprinting has been little studied. Actinomycin was found to inhibit imprinting by insulin[39], but did not inhibit imprinting by diiodotyrosine (T_2). Cycloheximide also inhibited imprinting by insulin, but puromycin had no influence on it[39].

Of the endocytosis inhibitors, colchicine and cytochalasin generally inhibit hormonal imprinting[39], except that colchicine does not interfere with imprinting by T_2[17].

The mechanism of hormonal imprinting seems to be highly sensitive to a great variety of external factors and internal changes. Hormonal imprinting of the target cell takes place only in a state of physiological integrity, which makes possible the mediation to the effector system of the signal which has arisen from receptor-hormone interaction, and subsequent recycling of it to the membrane as well. Dysfunction of any element of the effector system can account for the failure of hormonal imprinting.

Impact of the hormone concentration and time factor on hormonal imprinting

The applied hormone concentration and the duration of hormone exposure play a decisive role in hormonal imprinting.

A single treatment with T_2 increases the growth rate of the *Tetrahymena,* and this effect is concentration-dependent. A practically linear relationship has been demonstrated between the quantity of hormone and the growth rate of the cells in the concentration range 10^{-9}–10^{-15} M[18]. A single exposure to 10^{-18} M T_2 had in itself no influence on the growth rate, but a measurable increase occurred on reexposure at the same concentration, from which it follows that although the first exposure had not altered the growth rate, it did give rise to hormonal imprinting[22].

The growth rate showed a linear increase with the length of exposure from 10 min to 24 h, in cultures examined immediately after treatment, and a similar, although less distinct linearity was still demonstrable one week later. Reexposure to the hormone after one week did not significantly increase the growth rate of the cells preexposed for a short time (from 10 min to 1 h), but increased it significantly, if preexposure lasted 4 h or longer. Similar observations were made in cultures reexposed after two weeks[18].

It appears that although low hormone concentrations have no direct stimulatory effect on cell growth, they do induce imprinting, whereas high concentrations (10^{-9} M) have a direct stimulatory effect at short (1-h) exposure, but fail to induce imprinting. Thus the concentration of the hormone seems to play, at least within certain limits, a less important role in imprinting than the length of hormone exposure. If Koch's[29] dynamic receptor pattern generation theory (the continuous formation of membrane patterns from subpatterns, inquiring the milieu) is true, there is reason to postulate that the membrane-associated information carrier molecules require a certain time for assembling to a (receptor) pattern complementary to the signal molecule; hence, short-term exposures obviously cannot give rise to a durable imprinting.

In the case of insulin the optimal length of imprinting time (primary exposure) proved to be 24 h, when assessed 1 day after treatment, and 4 h, when assessed 28 days after it, although the cells preexposed for 4 h had bound less hormone than the control cells at 1 day after treatment. A 10-min primary exposure proved to be ineffective at both sampling intervals[36].

Higher than effective hormone concentrations (e.g. 10^{-4} M T_2), do not provoke hormonal imprinting[23].

Which materials can induce imprinting?

If the existence of imprinting is regarded as an established fact, the question may be justly posed, which types of molecules are capable of inducing the formation of binding sites in the membrane of the *Tetrahymena*?

From the point of view of hormone phylogenesis it should be taken into consideration that in many cases the *Tetrahymena* is more responsive to the phylogenetically lower hormone precursor(s) than to the hormone proper which represents the most active (signal) molecule in higher (e.g. mammalian) organisms. The classical example in this context is the thyroxine series (tyrosine,

T_1, T_2, T_3 and T_4), in which T_2 (diiodotyrosine) represents the most active signal molecule for the *Tetrahymena*[15]. In contrast, in the tryptophane series (tryptophane, tryptamine, 5-HTP and 5 HT) the most 'developed' molecule, 5-HT (serotonin) has the strongest action at the unicellular level. In other words, while the omnipresent hormone serotonin, which also occurs in the *Tetrahymena,* is universally most active in its proper form, the non-omnipresent vertebrate hormone thyroxine acts on the unicellular organism most strongly in its precursor (T_2) form[3, 15]. This phenomenon supports the implication that at lower levels of phylogenesis, the cell receptors also are phylogenetically lower structures than in higher organisms, although they appear to be less variable in their evolutionary course than the hormones[43].

The binding sites of insulin and Con-A overlap not only in mammalian cells, but also in the *Tetrahymena,* for they have common saccharide components at both phylogenetic levels. Several cellular responses elicited by insulin can also be provoked with Con-A. This prompted investigations into the imprinting capacity of Con-A for itself and for insulin. However, while insulin-imprinting accounted for a quantitative increase in membrane binding sites for Con-A (owing to common sugar components), Con-A-treatment failed to imprint the *Tetrahymena* for insulin, and imprinted it only temporarily for itself as well[36]. Non-hormone polypeptides, such as bovine serum albumin, protamine, etc. also failed to induce imprinting of similar duration to that induced by polypeptide hormones, such as insulin, glucagon, TSH and ACTH[16].

These observations strongly suggest that durable imprinting can only be induced by those signal molecules whose steric configuration enables them to act as a hormone; presumably molecules of this type have been transformed into hormones in the course of evolution[6].

Study of induced receptors in the Tetrahymena

It is important to clarify whether the receiver structures induced in the *Tetrahymena* by hormonal imprinting can be regarded as true receptors, and if they can, the following aspects remain to be investigated: 1) mode of receptor fixing and length of receptor persistence (demonstrability); 2) structural similarity to hormone receptors of higher organisms; 3) degree of receptor specificity, and hormone overlap phenomena on the receptor structure, of the kind observed in mammalian cells.

Investigations along these lines may throw light on similarities and dissimilarities between the hormone receptors of the *Tetrahymena* and those of higher organisms.

Receptor 'memory' in unicellular model systems

Diiodotyrosine (T_2) increases considerably the growth rate of the *Tetrahymena,* and the increase, although it tends to become less as time progresses, persists as long as 12 weeks, during which about 500 generation changes occur. Reexposure to T_2 elicits a still greater growth response, from which existence of a receptor 'memory'[20] and gene-level transmission of the information received can be deduced. Remarkably, repeated reexposures do not further increase the effect of the primary exposure (imprinting), but do not extinguish the 'memory' thereof.

Receptor 'memory' also operates in those cases in which the primary exposure (imprinting) does not in itself increase the growth rate over the control (e.g. in the case of epinephrine treatment), but the repeated treatment gives rise to a significant elevation of growth rate[19]. Induction of receptor 'memory' is not the privilege only of those hormones which are present in the body of the *Tetrahymena* (e.g. epinephrine), for 'foreign' hormones, such as the plant hormone gibberellin, can also evoke it[19].

Imprinting of the *Tetrahymena* with histamine gave rise to a measurable increase in the number of histamine receptors in as many as 50 subsequent generations, but reexposure to histamine 6 days after imprinting was followed by a significant decrease in histamine binding. In this particular case, receptor 'memory' operated for 'down-regulation'[11].

Certain similarities have been detected between receptor 'memory' and neuronal 'memory'. Four 1-h exposures to T_2 increased the growth rate of the *Tetrahymena* to a significantly greater degree than did a single uninterrupted exposure for 4 h[22]. This is still more remarkable, if it is taken into consideration that a single 1-h exposure could not induce hormonal imprinting in itself[18], but repeated short treatments evoked a long-lasting memory.

So-called retroactive interference was observed in those cases in which treatment with other hormones (serotonin, gramine, epinephrine, dopamine, or combinations of these) followed immediately after a 4-h primary exposure to T_2. In these conditions reexposure to a 'foreign' hormone depressed imprinting by T_2 or, more precisely, development of a receptor 'memory', to a considerable degree[22]. The only exception was serotonin, which depressed the growth stimulant action of T_2 only to a negligible degree, as assessed one week later. However, the 'foreign' hormones do not extinguish receptor 'memory' after it has become established, i.e. at a later time after preexposure. Membrane perturbations induced after the establishment of receptor 'memory' did not abolish imprinting either[46].

The theory of gene-level fixing of receptor 'memory', i.e. of hormonal imprinting (supported by the existence of 'memory' after 500 generations), is in good agreement with Robertson's[48] hypothesis that memory is a gene-level rather than membrane-level phenomenon.

In *Tetrahymena* cells once imprinted, 'memory' survives maintenance in anaerobic conditions, without medium exchange, for as long as a year. In these conditions the cells fail to divide, but their life span is extremely prolonged[8]. Investigations along this line may throw light on the true cause of decline of the memory with progressing time, i.e. whether mutation ('rejuvenation') of the cells in the course of serial divisions, or an inadequate coding of the new information, is responsible for the gradual loss of memory. At all events, the results of these experiments are inexplicable unless a gene-level fixing of the information is postulated, for the effect of imprinting by T_2 was still demonstrable after 9 to 12 months of growth arrest.

Important information has emerged from those experiments, in which the *Tetrahymena* cells were treated immediately after imprinting with materials structurally similar to, but functionally different from, the hormone used or, conversely, with others which were similar to the hormone in function, but differed from it in structure. Structural similarities had a more adverse influence on imprinting than functional ones[21], from which it follows that the conformation of the interacting molecule plays the leading role in hormonal imprinting.

Structural studies on membrane receptors of Tetrahymena

All receptor detection studies described in the foregoing sections were based on indirect approaches (assessment of growth rate, etc.), which do not supply adequate information on receptor structure. For structural studies, immunological tests are the method of choice.

In experimental conditions, rat hepatocellular insulin receptors provoke the formation of specific antibodies, which bind to insulin-imprinted *Tetrahymena* cells to a significantly greater degree than to intact control cells, and induce the formation of additional insulin-binding sites in the membrane of the *Tetrahymena*. It appears that the induced insulin binding sites of the *Tetrahymena* are immunologically similar to the insulin receptors of the rat hepatocellular membrane[13].

Rabbits immunized with live *Tetrahymena* cells develop antibodies presumably to the surface components of the cells, which also include insulin receptors of the *Tetrahymena* cells used for immunization had previously been exposed to insulin. Treatment of rat hepatocytes with antibodies to *Tetrahymena* was followed by a decrease in their insulin binding capacity. Antisera prepared with insulin-treated *Tetrahymena* cells accounted for a significantly greater decrease than those prepared with intact cells. FITC-labeled antiserum to insulin-treated *Tetrahymena* increased hepatocellular insulin binding capacity to a significantly greater degree in insulin-treated than in untreated control cells[35]. These phenomena, too, suggest a structural similarity between the induced insulin binding sites of the *Tetrahymena* and the insulin receptors of rat hepatocytes. It also follows from the above experimental observations that the insulin receptors of the *Tetrahymena* are not preformed structures, but arise under the influence of the hormone, since hormonal presence is an indispensable prerequisite of receptor formation.

Specificity of the hormone receptors of Tetrahymena

Experimental studies have unequivocally shown that the receptors of *Tetrahymena* are highly specific. The H_1- and H_2-type histamine receptors of the unicellular organism have dissimilar saccharide components[38]. The chemically histamine-like histamine antagonists bind to the same sites as histamine itself, while the structurally dissimilar ones bind either to other sites, or to those regions of the histamine receptor which do not interfere with histamine binding but prevent histamine action. Obviously, a marked specialization of binding sites exists also at the lowest levels of phylogenesis.

In mammalian cells, structurally related hormones (e.g. TSH and FSH) overlap on each other's binding sites, so that imprinting with TSH amplifies the receptor not only for itself, but also for FSH, and vice versa[26]. Such overlaps are also expressed in physiological responses[9]. TSH-FSH and FSH-TSH overlaps have also been demonstrated in *Tetrahymena* cells[32], although in these only TSH amplified for both itself and FSH, whereas FSH amplified to a lesser degree for TSH than for itself. Similar phenomena were observed in mammalian cell cultures, too[26]. The differences between the imprinting actions of the two hormones indicate that hormone overlaps are not greater at the unicellular than at the mammalian level, and the *Tetrahymena* cell, too, is capable of differentiating structurally related hormones from one another.

Experiments with serotonin and other structurally related hormones (5-HTP, 5-methoxytroptophane, tryptamine, gramine, indoleacetic acid) have also presented evidence in support of receptor specificity. There are indications that the receptor recognizes the basic structure of the signal molecules. Presence or absence of a hydroxyl group had little influence on hormone-receptor interaction, but the presence of two methyl groups practically excluded any cellular response[14]. Presence of a methoxy- or hydroxyl group in position 5 of serotonin and related molecules increases binding affinity for the receptor, as has been observed earlier in mammalian cells[27].

Summarizing, we may state that *Tetrahymena* is an ideal model for investigations into the mechanism of hormonal imprinting. Experiments on that model can provide important information on the course of imprinting, and the properties, structure, 'memory' and specificity of the induced receptors can be relatively easily studied by an appropriate experimental approach.

It appears that, although there are certain essential differences between the response of unicellular *(Tetrahymena)* and higher (e.g. mammalian) organisms to hormones, hormonal imprinting is in many respects similar at the low and high levels of phylogenesis. Similarities between *Tetrahymena* and mammalian cells include the immunologically-studied nature of the hormone receptors; the overlaps of lectin and hormone binding sites; and the critical stages of the imprinting mechanism. In view of this, the information emerging from receptor studies on the *Tetrahymena* can be – with great caution – extrapolated to the cells of higher organisms. Experimental observations on receptor formation in the *Tetrahymena* are in good agreement with current conceptions on the phylogeny and ontogeny of receptors[2-4, 6, 7], and unequivocally support the hypothesis that phylogenetically, receptor development had its origin in the imprinting mechanism, which is reproduced during ontogenesis.

1 Birnbaumer, L., The actions of hormones and nucleotides on membrane-bound adenyiyl cyclases: an overview, in: Receptors and Hormone Actions, vol. 1, pp. 485–547. Eds W. O. O'Malley and L. Birnbaumer. Academic Press, New York, San Francisco, London 1977.

2 Csaba, G., Phylogeny and ontogeny of hormone receptors: the selection theory of receptor formation and hormonal imprinting. Biol. Rev. *55* (1980) 47–63.

3 Csaba, G., Ontogeny and Phylogeny of Hormone Receptors. Monogr. devl. Biol, vol. 15. Karger, Basel 1981.

4 Csaba, G., The present state in the phylogeny and ontogeny of hormone receptors. Horm. Metab. Res. *16* (1984) 329–335.

5 Csaba, G., The unicellular *Tetrahymena* as model cell for receptor research. Int. Rev. Cytol. *95* (1985) 327–377.

6 Csaba, G., Why hormone receptors arise? Experientia *42* (1986) 715–718.

7   Csaba, G., Receptor ontogeny and hormonal imprinting. Experientia *42* (1985) 750–759.
8   Csaba, G., Darvas, Zs., László, V., and Vargha, P., Influence of prolonged life span on receptor 'memory' in a unicellular organism, Tetrahymena. Expl Cell Biol. *52* (1984) 211–216.
9   Csaba, G., Dobozy, O., and Kaizer, G., FSH-TSH functional overlap in cockerel testicle. Durable amplification of the hormone receptor by treatment at hatching. Horm. Metab. Res. *13* (1981) 177–179.
10 Csaba, G., and Kovács, P., Histamine-lectin and insulin-lectin binding site overlaps in Tetrahymena. Cell. molec. Biol. *28* (1982) 153–158.
11 Csaba, G., and Kovács, P., Further experiments in unicellular model system to substantiate receptor amplification. Endokrinologie *79* (1982) 242–246.
12 Csaba, G., and Kovács, P., The role of second messengers in the hormonal imprinting. II. The role of cAMP in the hormonal imprinting in Tetrahymena. Acta physiol. hung. in press (1986).
13 Csaba, G., Kovács, P., and Inczefi-Gonda, A., Insulin binding sites induced in the *Tetrahymena* by rat liver receptor antibody. Z. Naturf. *396* (1984) 183–185.
14 Csaba, G., and Lantos, T., Specificity of hormone receptors in Tetrahymena. Experiments with serotonin and histamine antagonists. Cytobiologie *11* (1975) 410–413.
15 Csaba, G., and Németh, G., Effect of hormones and their precursors on protozoa – the selective responsiveness of Tetrahymena. Comp. Biochem. Physiol. *65B* (1980) 387–390.
16 Csaba, G., Németh, G., Kovács, P., and Vas, Á., Receptor level action of polypeptide hormones (insulin, glucagon, TSH, PMSG, ACTH) and non-hormone polypeptides on Tetrahymena. BioSystems *17* (1985) 227–231.
17 Csaba, G., Németh, G., and Vargha, P., Experimental observations on the mechanism of hormonal imprinting: Influence of actinomycin D, methylamine and colchicine on receptor memory in a unicellular model system. Endokrinologie *80* (1982) 341–346.
18 Csaba, G., Németh, G., and Vargha, P., Influence of hormone concentration and time factor on development of receptor memory in a unicellular (Tetrahymena) model system. Comp. Biochem. Physiol. *73B* (1982) 357–360.
19 Csaba, G., Németh, G., and Vargha, P., 'Memory' of first interaction with physiological or biologically active foreign molecules (benzpyrene, gibberelline) in a unicellular *(Tetrahymena)* model system. Z. Naturf. *37c* (1982) 1042–1044.
20 Csaba, G., Németh, G., and Vargha, P., Development and persistence of receptor 'memory' in a unicellular model system. Expl Cell Biol. *50* (1982) 291–294.
21 Csaba, G., Németh, G., and Vargha, P., Attempt to disturb receptor memory in a unicellular *(Tetrahymena)* model system. Acta physiol. hung. *61* (1983) 131–136.
22 Csaba, G., Németh, G., and Vargha, P., Receptor 'memory' in Tetrahymena: Does it satisfy the general criteria of memory? Expl Cell Biol. *52* (1984) 320–325.
23 Csaba, G., Németh, G., and Vargha, P., Influence of single or multiple treatments by related low or high concentration substances on the diiodotyrosine stimulated growth of Tetrahymena. Biológia *32* (1984) 73–76.
24 Csaba, G., and Sudár, F., Localization of ^{3}H-histamine in the nucleus of Tetrahymena. Acta morph. hung. *27* (1979) 89–94.
25 Csaba, G., Sudár, F., Nagy, S.U., and Dobozy, O., Localization of hormone receptors in Tetrahymena. Protoplasma *91* (1977) 179–189.
26 Csaba, G., Török, O., and Kovács, P., Hormonal imprinting in cell culture. II. Evidence of hormonal imprinting and thyrotropin (TSH)-gonadotropin (FSH) overlap in chinese hamster ovary (CHO) cell line. Acta physiol. hung. *64* (1984) 135–138.
27 Gleman., R.H., and Gessner, P.K., Serotonin and receptor binding affinities of tryptamine analogues. J. med. Chem. *22* (1979) 428–432.
28 Gundersen, R.E., and Thompson, G.A. Jr, Factors influencing the pattern of dopamine secretion in Tetrahymena pyriformis. Biochim. biophys. Acta *755* (1983) 186–194.
29 Koch, A. SA., Fehér, J., and Lukovits, I., A single model of dynamic receptor pattern generation. Biol. Cybernet. *32* (1979) 125-138.
30 Kovács, P., Influence of environmental (culturing) conditions on the lectin-binding capacity of Tetrahymena. Acta biol. hung. *35* (1984) 83–90.
31 Kovács, P., and Csaba, G., Detection of histamine binding sites (receptors) in Tetrahymena by fluorescence technique. Acta biol. med. germ. *39* (1980) 237–241.
32 Kovács, P., and Csaba, G., Receptor level study of polypeptide hormone (insulin, TSH, FSH) imprinting and overlap in Tetrahymena. Acta protozool. *24* (1984) 37–40.
33 Kovács, P., and Csaba, G., Is the saccaride component of the insulin receptor involved in hormonal imprinting? Cell Biochem. Funct. in press (1986).

34 Kovács, P., and Csaba, G., The role of second messengers in the hormonal imprinting. I. The role of Ca^{2+} in hormonal imprinting in Tetrahymena. Acta physiol. hung., in press (1986).

35 Kovács, P., Csaba, G., and Bohdaneczky, E., Immunological evidence of the induced insulin receptor in Tetrahymena. Comp. Biochem. Physiol. *80A* (1985) 41–42.

36 Kovács, P., Csaba, G., and László, V., Study of the imprinting and overlap of insulin and concanavalin A at the receptor level in a protozoan *(Tetrahymena)* model system. Acta physiol. hung. *64* (1984) 19–23.

37 Kovács, P., Csaba, G., and Nozawa, Y., Influence of membrane fluidity changes upon the imprinting of polypeptide hormones in Tetrahymena. Comp. Biochem. Physiol. *78A* (1984) 763–766.

38 Kovács, P., Darvas, Zs., and Csaba, G., Investigation of histamine-antihistamine differentiation ability of Tetrahymena receptors, by means of lectins and antihistamine antibodies. Acta biol. hung. *32* (1981) 111–117.

39 Köhidai, L., Kovács, P., and Csaba, G., Effects of inhibitors of protein synthesis and endocytosis on hormonal imprinting in the Tetrahymena. Acta protozool. *24* (1984) 259–264.

40 Lenhoff, H. M., Behavior, hormones and hydra. Science *161* (1968) 432–442.

41 Le Roith, D., Shiloach, J., Berelowitz, M., Frohman, L. A., Liotta, A. S., Krieger, D. T., and Roth, J., Are messenger molecules in microbes the ancestors of the vertebrate hormones and tissues factors? Fedn Proc. *42* (1983) 2602–2607.

42 Le Roith, D., Shiloach, J., Roth, J., and Lesniak, M. A., Evolutionary origins of vertebrate hormones: substances similar to mammalian insulins are native to unicellular eukaryotes. Proc. natn. Acad. Sci. USA *77* (1980) 6184–6188.

43 Muggeo, M., Ginsberg, B. H., Roth, J., Neville, O. M., De Meyts, P., and Kahn, C. R., The insulin receptor in invertebrates is functionally more conserved during evolution than the insulin itself. Endocrinology *104* (1979) 1393–1402.

44 Nakamura, Y., Maekawa, Y., Katayama, S., Okada, Y., Szuki, F., and Nagata, Y., Induction of a metallothienin-like protein in *Tetrahymena* pyriformis by metal ions. Agric. biol. Chem. *45* (1981) 2375–2377.

45 Nozawa, Y., Kovács, P., and Csaba, G., The effects of membrane perturbants, local anesthetics and phenothiazines on hormonal imprinting in Tetrahymena pyriformis. Cell. molec. Biol. *31* (1985) 223–227.

46 Nozawa, Y., Kovács, P., and Csaba, G., Influence of membrane disturbances elicited after hormon treatments (hormonal imprinting) on the later hormone binding capacity of Tetrahymena. Cell. molec. Biol. *31* (1984) 13–16.

47 Nozawa, Y., Kovács, P., Csaba, G., Ohki, K., Effect of internalization of insulin-encapsulated and empty liposomes on hormone binding and its imprinting in Tetrahymena. Cell. molec. Biol. *31* (1984) 7–11.

48 Robertson, M., Learning, forgetting and the cell biology of memory. Nature *300* (1982) 219–220.

49 Schlatz, L., and Marienetti, G. V., Hormone-calcium interactions with the plasma membranes of rat liver cells. Science *176* (1972) 175–177.

50 Schultz, J. E., Schönfeld, W., and Klumpp, S., Calcium/calmodulin-regulated guanylate cyclase and calcium-permeability in the ciliary membrane from Tetrahymena. Eur. J. Biochem. *137* (1983) 89–94.

51 Sorimachi, K., and Yasumura, Y., Concanavalin A can trap insulin and increase insulin internalization into cells cultured in monolayer. Biochem. biophys. Res. Commun. *122* (1984) 204–211.

52 Suzuki, T., Makino, H., Kanatsuka, A., Osegawa, M., Yoshida, S., and Sakamoto, Y., Activation of insulin-sensitive phosphodiesterase by lectins and insulin-dextran complex in rat fat cells. Metabolism *33* (1984) 572–576.

53 Umeki, S., Maruyama, H., and Nozawa, Y., Studies on thermal adaptation of *Tetrahymena* lipids. Alteration in fatty acid composition and its mechanism in the growth temperature shift-up. Biochim. biophys. Acta *752* (1983) 30–37.

The phylogeny of the endocrine system

E. J. W. Barrington

formerly University of Nottingham; 2 St. Margarets Drive, Alderton, Tewkesbury, Glocestershire GL208NY (England)

The scope of phylogeny

Phylogeny, which is a representation of the evolutionary history of taxa, requires critical interpretation of biological diversity and of the relationship of this with the unity of organisation which underlies living systems. Some degree of personal judgement may be involved in particular cases, and so, because of this subjective element, phylogenetic propositions are always open to discussion in the light of new information. On the larger issues, however, and on many smaller ones as well, there is substantial agreement, founded on the rich classical resources of descriptive anatomy, which take account of large numbers of species, both living and extinct, and with embryological studies as a valuable supplement. The strength of these views derives not only from the wide range of material studied, but also from the possibility of evaluating the anatomical interrelationships of organs and systems which may be unduly influenced by convergence[10]. Further evidence is, of course, provided by physiological, biochemical and molecular studies, but in the euphoria engendered by recent advances in some of these areas it is easy to overlook their limitations: the paucity of species studied, ignorance of extinct forms, and the flexibility of functional organisation and its consequent susceptibility to opportunistic and convergent evolution. Caution is therefore demanded.

Nevertheless, it is fair to say that current views of the phylogeny of endocrine systems do conform in general with classical views of evolutionary history, yet the correspondence is not absolute, and our expectations are not always satisfied. For example, sequence studies[28] of the insulins of the chick, an alligator and two species of snakes support classical phylogenetic deductions drawn from living and fossil vertebrates, but it is salutary to be told that hormones do not always show such clear patterns; snakes, for example, possess an unusual gonadotropin, and perhaps only a single one[31].

However, limitations inherent in interpretations of endocrine phylogeny do not reduce them to a sterile pursuit. On the contrary, phylogenetic perspectives are essential if we are to analyse in sufficient depth such fundamental issues as the origin and diversification of biologically active molecules and their receptors, the exploitation of their hormonal potentialities, and their resulting contributions to complex and adaptively flexible regulatory systems. A brief review of some aspects of thyroid phylogeny will illustrate this argument.

Thyroidal phylogenesis

Thyroidal biosynthesis involves the uptake of iodide, its oxidation to reactive iodine by peroxidase, and the binding of this within thyroglobulin molecules into iodotyrosines which are then polymerised to form thyroxine and triiodothyronine. The gland is sharply restricted to vertebrates, yet organic binding of iodine, demonstrable by autoradiography, occurs in a number of invertebrates, in some of which small amounts of iodotyrosines and, less commonly, iodothyronines have been found. However, such binding is often associated with the formation of structural proteins, so that these iodinated products cannot be circulated and made available for general metabolism[2]. What, then, were the special circumstances that permitted the establishment of a thyroid gland so early in vertebrate phylogeny that it could be already well developed in the primitive jawless Agnatha (lampreys and hagfish)?

The generally accepted explanation is that the thyroid originated in a pharyngeal gland, the endostyle. This forms part of a unique ciliary feeding mechanism found in the protochordates (e.g. amphioxus and the ascidians), which are surviving representatives of early forerunners of the vertebrates. Peroxidase[13, 27] and a thyroglobulin-like material[49] have been identified in the endostyle. So also have iodotyrosines and at least one iodothyronine[5], which can be absorbed in the intestine from the endostylar secretion, and thus be metabolically exploited. On this argument, phylogenetically ancient molecules have been introduced into the vertebrate endocrine system as a consequence of their production under uniquely favourable conditions in remote ancestors. It makes a plausible phylogenetic statement, but can we identify selective advantages that might have determined this proposed course of events?

The structural features required for thyroidal activity include 3, 5, 3′ substituents which need not be iodine or, indeed, halogens at all; halogen-free methyltyrosines, for example, have significant thyromimetic activity in tadpoles and rats. What, then, were the selective advantages of the iodothyronines? Frieden[17] points out that they would presumably have included the ready availability of iodine in the sea, where the chordate line arose. This advantage was, of course, lost when vertebrates passed through fresh water to the land, but by then it must have been too late to abandon an established molecular relationship. Other circumstances favouring thyroidal evolution would have included the availability of H_2O_2-peroxidase systems, and of ample proteases for the release of the iodothyronines from their peptide linkages, together with the possibility of re-utilising iodine after their deiodination. The presence of thyroglobulin-like molecules in the endostyle may well have favoured molecular orientation suitable for the polymerisation of the iodotyrosines.

But even with these advantages, thyroidal biosynthesis could hardly have become established unless the iodothyronines had some survival value, and it is difficult to suggest what this might initially have been. Indeed, it has been well said of the whole field of thyroid action that 'a huge and bewildering array of observations ... has generated more confusion than clarification'[17]. There are gaps, therefore, in the phylogenetic argument, and it is important to recognise them, for they are not peculiar to thyroid phylogeny. It has, however, been suggested that thyroidal biosynthesis might have served at first to sequester and bind iodine for general metabolic use[16], and that only later did the iodothyronines become involved in regulating growth, reproduction and metamorphosis, as they do in the lower vertebrates. The evolution of their calorigenic function in the higher forms was probably correlated with the establishment of homeothermy; responses of the gland to high temperature, which have been detected in lower vertebrates, may have provided the starting point. But these arguments remain a speculative contribution. They are part of 'the vastness of the unexplored realm' of endocrine phylogeny[16].

The steroid ring system

The iodothyronines are an example of molecules with latent hormonal potential, that had a wide distribution and long record before that potential came to fruition. Steroids, however, provide a much more ample illustration of this. Omnipresent throughout the biosphere, and with marked capacity for molecular diversification, they have been speculatively viewed as 'very ancient bioregulators' which evolved prior to the appearance of eukaryotes[45]. Their use as hormones is well established in two phylogenetically unrelated lines: the arthropods, where ecdysteroids contribute to the regulation of moulting, and the vertebrates, where androgens and oestrogens regulate sexual characteristics and reproduction, while adrenocorticosteroids contribute to the maintenance of homeostasis.

The strong conservation of these vertebrate hormones (some of which are already present in agnathans) and of their biosynthetic pathways has led to the conclusion[45] that 'early hopes of establishing evolutionary tendencies' have not been realised, and that 'class divergences are due to environmental adaptation and not to evolution'. But some taxonomic diversification is discernible and, of course, environmental adaptation *is* evolution in action, revealing the power of selection to mould the individuals which are its targets[33]. The difficulty, as with the analysis of thyroid phylogeny, is that adaptive significance is not always easy to see.

One well-defined taxonomic feature is the presence of 1α-hydroxy-corticosterone in elasmobranch fish, where it probably acts as a mineralocorticoid. The existence in vertebrates of a 1α-hydroxylated derivative of vitamin D perhaps accounts for its origin[45]. Selective advantage can, however, be ascribed to the emergence of aldosterone as a mineralocorticoid in lungfish and tetrapods. Adaptation to terrestrial life may well have favoured a sharper separation of glucocorticoid action from mineralocorticoid, the requirement for this being a molecule sufficiently distinctive to be recognised by separate receptors. A further advantage of the separation could have been that it made possible some more precise action of the controlling mechanisms, with cortisol and corticosterone being regulated through the pituitary, and aldosterone mainly by the renin-angiotensin system[6].

Although these and other steroid molecules are highly characteristic of the vertebrate endocrine system, some of them occur widely in invertebrates and also in plants. Thus testosterone and progesterone are present in the haemolymph of the fleshfly, *Sarcophaga*[12], and testosterone in the serum and testes of the American lobster, *Homarus americanus*[8], while the in vitro conversion of cholesterol to vertebrate-type steroid molecules has been reported for the spiny lobster, *Panulirus japonica*[25]. Progesterone is among those reported in plants. Whether these molecules have regulatory functions in invertebrates is still matter for speculation. As for plants, Heffmann[22], while accepting that the significance of the existence in them of vertebrate type steroids is unknown, has argued for the possibility that they may interact with plant chromosomes as they do with those of mammals. The evidence for this is, to say the least, tenuous, but it could well be that such molecules, both in plants and in invertebrates, have functions unrelated to chemical communication. Certainly it would be unwise to assume at this stage that they have no functions at all. As for the vertebrates, we can only draw the obvious conclusion that they have exploited the endocrine potentialities of widely distributed molecules, the functions of which in other groups remain obscure.

Peptide hormones

It is particularly in the field of peptide hormones that the most impressive advances have been made in our approach to endocrine phylogeny. Already they have amply confirmed, as was already apparent from thyroid and steroid phylogeny, that the boundaries traditionally set for the phyletic distribution of regulatory molecules have been far too narrow[43]. Developments in peptide chemistry have led to the precise characterisation of the molecules, while related techniques such as autoradiography, immunoassay, and immunocytochemistry have made possible the measurements of minute quantities of circulating hormones and the location of their sites of origin and of action. Such is the wealth and complexity of the continuously accumulating data that it will not be possible to do more here than indicate some emerging principles that bear upon phylogenetic issues.

One important advantage of peptide hormones, in the context of endocrine phylogeny, is that they give direct insight into the genetic mechanisms that generate adaptive diversification. Novelties can, in principle, arise at the level of the DNA code by point mutations bringing about amino acid replacements in peptide sequences, but the evolutionary possibilities of this are limited by the need to maintain the established properties of the molecules concerned. It is to be expected that strong selection pressure will conserve molecular structure, and thus ensure the stability of binding sites and other functionally important regions.

For example, insulin molecules from different species differ in respect of a number of substitutions, yet the agnathan (hagfish) insulin differs from that of man in only about 38% of its residues[35]. It is impressive to find such stability after some 500 million years of independent agnathan and gnathostome evolution.

Those substitutions that have occurred in the insulin molecule are predominantly conservative, which is why the variants share common biological properties[51]. Why, then, have these substitutions become established? Are they the result of random change, or are they adaptively significant? There is much to support the latter view, for neither internal nor external environments are static. Point mutations in biologically active molecules may therefore provide the fine-tuning needed to ensure their continuing efficiency in changing conditions[4]. Certainly in so complex a problem it would be unwise to assume that mutations with no immediately obvious function are necessarily non-adaptive. A more prudent view is that 'the concept that neutral mutations can account for a large proportion of sequence variations in proteins looks increasingly unattractive'[21]. It has also been argued that the individual is the ultimate target of selection; any neutral mutations that it may carry are therefore an irrelevant issue, and are 'merely hitch-hikers of successful genotypes'[33].

Interspecific variation is, of course, common in vertebrate peptide hormones, and has been reported also in the invertebrates. Thus extraction and characterisation of potent hyperglycaemic hormones from several crustaceans[32], including an isopod *(Porcellio)* and two decapods *(Carcinus* and *Orconectes)* have shown them to be peptides with 50–58 residues, with overall similarity of composition, but with much interspecific variation in detail. Cross-reaction studies have shown in this case that the receptors have varied side by side with the hormones.

Despite the limitations restricting the evolutionary influence of point mutations in established molecules, some functional diversification is not precluded, particularly when associated with taxonomic separation sufficient to favour the establishment of new target relationships. This is well exemplified in invertebrates by the resemblance between the 8-residue erythrophore-concentrating hormone of crustaceans (ECH) and the 10-residue adipokinetic hormone (AKH) of insects. These two hormones, influencing respectively pigment cells and fat metabolism, are sufficiently alike structurally to imply divergence by point mutations from a single ancestral molecule[34]. With them can be grouped two further peptides, periplanetin CC-1 and periplanetin CC-2, which both show adipokinetic activity in grasshoppers and hyperglycaemic activity in cockroaches[4].

ECH Glu-Leu-Asn-Phe-Ser-Pro-Gly-Trp-NH$_2$
AKH Glu-Leu-Asn-Phe-Thr-Pro-Asn-Trp-Gly-
     Thr-NH$_2$
CC1 Glu-Val-Asn-Phe-Ser-Pro-Asn-Trp-NH$_2$
CC2 Glu-Leu-Thr-Phe-Thr-Ser-Asp-Trp-NH$_2$

Here, then, is a family of four peptides derivable in principle by point mutations from a common ancestral molecule, with some functional diversification.

Pathways of endocrine diversification

The evolutionary potentialities of point mutations become much greater when they are associated with gene duplication. Indications of repeated sequences (termed internal homologies) in large molecules (somatotropin is a case in point) suggest that they may have evolved through the association of the products of such duplication. But separation of the products has still greater potentiality, for this makes possible the continued functioning of the already established molecule while leaving the other product free to accept a range of mutations and thus to become available for new functions.

This is particularly well exemplified by two mammalian pituitary hormones: somatotropin (growth hormone), with 189–191 residues, and prolactin, with 199. Sequence studies have shown the ovine hormones to have some 23% of their residues in common, which is generally conceded to indicate divergence from a common molecular ancestry. A third member of the family, human placental lactogen, which shares some 85% of its residues with humor somatotropin, must have diverged much later, probably during primate evolution, with consequently much less time for mutations to accumulate[52]. The divergence of somatotropin and prolactin, however, must have occurred very early in vertebrate evolution, for both are present, on immunological evidence, in the agnathan (lamprey's) pituitary[53]. As a result, prolactin has had a long period of independent evolution during which it has established a range of target relationships so wide that it is difficult to attribute any one major function to it. In general, though, it seems to be particularly associated with transport-regulating effects[7].

It must suffice to mention one other example of gene duplication, with subsequent mutation, as a source of diversification. This concerns the dual series of neurohypophysial nonapeptides, represented in placental mammals by oxytocin

Cys-Tyr-Ile-Glu-Asn-Cys-Pro-Leu-Gly-(NH$_2$)
  1   2   3   4   5   6   7   8   9

and either arginine vasopressin (Tyr^2, Phe^3, Arg^8) or, in the pig and related forms, by lysine vasopressin (Tyr^2, Phe^3, Lys^8). Because of the similarity of the two series, and because only one peptide (vasotocin: Tyr^2, Ile^3, Arg^8) is found in agnathans, it is commonly assumed that the dual series arose through gene duplication during the emergence of gnathostomes[1]. Thereafter, some taxonomic diversification occurred by amino acid substitution in the several vertebrate classes, but this is not easy to correlate with selective pressure, except for the replacement of vasotocin by vasopressin at the origin of mammals, for the new molecule has the advantage of much greater antidiuretic activity.

Further evidence of gene duplication in this molecular family is found in the marsupials, some of which have been shown to possess two members of the vasopressin series[4]. Arginine and lysine vasopressin occur in two species of South American opossums, while lysine vasopressin and a previously unrecognised molecule, phenypressin (Phe^2, Phe^3, Arg^8) occur in five Australian macropodid species. It can be assumed that gene duplication of the vasopressin gene occurred in marsupial ancestors, followed by independent mutation in the South American and Australian continents, but it is not yet possible to relate these postulated events to any form of selection pressure[9]. Further data, however, may be expected to clarify the problems presented by this remarkable phylogenetic series.

It remains to mention one other pathway for genetic diversification. This is post-translational processing and cleavage of large precursor molecules. A remarkable illustration of this is the production of a number of molecules, including several pituitary hormones, from proopiocortin[37], a single large

(37,000 mol.wt) polypeptide. In the corticotroph cells of the mammalian pituitary it yields corticotropin and β-LPH (β-lipotropin) as final products, but in
the melanotroph cells the corticotropin is further cleaved to yield α-MSH
(melanocyte-stimulating hormone) and CLIP (corticotropin-like intermediate
lobe peptide), while β-LPH yields β-MSH, β-endorphin and another fragment.
This sequence of events seems to have been established very early in vertebrate
phylogeny, for the precursor is present in the salmon, but with some distinctive
characteristics[26]. It is thought also to be present in agnathans, although it may
there be appreciably different from that of gnathostomes[38]. The origin of the
precursor is unknown, as also is the history of its products, but the presence in it
of repeated MSH-like sequences suggests that, as with somatotropin, gene
duplication, with association of the products, may have played some part.
Leaving aside many matters of detail which are outside the scope of this
account, one cannot fail to be impressed with the remarkable flexibility of the
genetic mechanism in providing a basis for the emergence of novelty and the
establishment of major adaptation, while also ensuring molecular stability and
any fine-tuning needed for functional efficiency. Acher[1] emphasises the complexity of these processes, and the many places at which variations can occur.
The peptide molecule must be produced in a stable form adequate to fit its
receptors, while the conformation of the latter is itself the result of evolution.
Add to this the specialisation of the target cell, and one can appreciate the
complex chains of intermediate molecules and enzymes which have to be integrated to secure adaptively valuable results. Natural selection is the only known
mechanism which can integrate so many initially disparate factors into an
orderly phylogenetic sequence.

Multiple sites and actions

New insights into the increasingly complex area of endocrine phylogeny have
been provided by evidence, primarily but not entirely immunocytochemical,
that many known peptide hormones of vertebrates may also be present in
identical or closely related forms in the nervous system, and particularly in the
brain[1]. The alimentary hormone cholecystokinin (CCK) is one example, with 33
residues in the alimentary mucosa, where it was first identified, but mainly with
8 in the nervous system. The difference may be attributed to differential processing of a macromolecular precursor. Conversely, somatostatin, first identified as
a factor transmitted in the hypophysial portal system to inhibit the release of
somatotropin, was later identified in alimentary sites, notably in the D cells of
pancreatic islet tissue, where it is thought to inhibit the release of insulin from
the B (insulin-secreting) cells by local (paracrine) action[51]. It must be remembered that antigenic determinants may comprise only a few amino acids, and
that the sequence involved may not include that part of the molecule responsible
for biological activity. Nevertheless, the evidence is already sufficiently powerful to justify current viewing of the nervous system as a secretory centre of great
complexity.
With the wisdom of hindsight, one can see this as a corollary of the existence of
neurosecretory cells. These, according to an earlier operational definition[48],

were distinguished from conventional neurons because they discharged chemical mediators (neurohormones) that were conveyed in the blood stream to function at a distance like the hormones produced by epithelial endocrine glands. These neurohormones were later shown to be peptides; the cells secreting them were therefore termed peptidergic, distinguishable from the cholinergic or aminergic conventional neurons. However, the great variety of chemical signals now known to be released in the vertebrate central nervous system has weakened this distinction. It has been suggested instead[24] that two types of transmission might be recognised: 'anatomically addressed' ones, involving point-to-point transmission in the central nervous system, and mediated principally by γ-aminobutyric acid and L-glutamate, and 'chemically addressed' ones, dependent upon the monoamines and the many neuropeptides, of which over 30 have now been identified. It is proposed that 'chemically addressed' systems are characterised by a rich diversity of signals, a slower time course, and less precise connections than are provided by the 'anatomically addressed' ones[24]. Be this as it may, the brain is certainly emerging as a vastly complicated neuroendocrine secretory organ. Associated with this view is recognition that neural signalling and endocrine regulation are not two distinct activities with independent origins, but rather as parts of a regulatory complex from which the endocrine system sensu stricto separated and then followed its own course. To examine the fuller implications of this approach it will be necessary briefly to consider the situation in invertebrates.

Invertebrates

That invertebrates are likely to have much in common with the vertebrates in these respects is suggested by immunocytochemical studies of protochordates which, as we have seen, take us back close to the remote ancestors of vertebrates. Positive results have been obtained for a wide range of vertebrate-like peptide material in the alimentary tract and cerebral ganglia of the ascidian Ciona[18]. Cells in the alimentary epithelium of Branchiostoma (amphioxus) react to antisera against a number of mammalian peptides, including insulin, glucagon, pancreatic polypeptide, somatostatin, secretin, vasoactive intestinal polypeptide, pentagastrin and neurotensin, and show specific distribution patterns[42]. Further, the brain has neurosecretory systems of surprising complexity for so small an animal. Monoaminergic and peptidergic neurons have been identified, while a central neurohaemal area is thought to be comparable to the median eminence and neurohypophysis of vertebrates[39, 40].

Amongst other (non-chordate) invertebrates it is the insects that have been most closely studied, and here, although conclusive demonstration of the neurotransmitter roles of suspected agents are still needed, there is ample evidence that many vertebrate neurotransmitters are also active in these animals and in other invertebrates as well[23]. Substances thought to act as neurotransmitters in insects include acetylcholine (predominant in the central nervous system) and glutamic acid (at excitatory neuromuscular junctions), with monoamines acting at various sites. Neurosecretory peptides are present throughout the central nervous system of these animals (bursicon and eclosion hormones being examples characteristic of the group), but there is also immunological evidence of a range of substances resembling such typical vertebrate products as somatostatin, insulin, glucagon and gastrin.

Comparisons with vertebrates need, however, to be drawn with caution. For example, immunoreactive somatostatin has been found in two species of pond snail (*Lymnaea stagnalis* and *Physa* spp.) and has been thought to be a growth factor[19], but it seems not to be chemically identical to synthetic somatostatin. Again, the ganglia of the mollusc *Aplysia* and of two slugs contain a neurohypophysial peptide-like material which resembles vasotocin and vasopressin, but is actually neither. It is suggested[46], although quite hypothetically, that this material might act as a neurotransmitter or neurohormone in the regulation of fluid balance.

These data, and much else besides, are confirming the view that the phylogenetic history of vertebrate hormones must extend outside the group. The strong indications of insulin-like material in a number of invertebrates provide a good illustration of this, studies of insects being particularly convincing. Extracts of the brain of the blowfly *(Calliphora)* contain a substance similar to insulin in physicochemical properties as well as in biological activities[14], while insulin B-chain immunoreactivity has been located in a few cells in the frontal ganglion of the tobacco hornworm *(Manduca)*[15]. A complication still awaiting interpretation in terms of molecular evolution, however, is the demonstration in the adult silkworm moth *(Bombyx)* of three forms of a well-defined insect hormone (prothoracotropic hormone) and of a significant resemblance between these and the A-chain of insulin[36].

The possible functions of insulin-like materials in invertebrates remain to be discovered. They have been thought to be involved in the regulation of carbohydrate metabolism in bivalve molluscs[41], but probably to have no glucostatic role in the lobster *Homarus americanus*[44]. Of course, the functions of endocrine molecules need not always be the same, regardless of the group in which they occur. A case in point is the detection in the central nervous system and haemolymph of the pond snail *(Lymnaea stagnalis)* of an immunological response resembling that of the thyrotropin-releasing hormone (TRH) of the vertebrate pituitary. Obviously the action of this substance, whatever it may be, cannot be that of the vertebrate hormone, which is transmitted in the hypophysial portal system to evoke release of thyrotropin. There is, however, experimental evidence that exogenous TRH may influence hydromineral balance in gastropod molluscs by an action on the secretion of sulphated polysaccharides by the epithelium of the foot[20]. Too little attention is sometimes given to the profound differences in organisation and mode of life when comparisons are made between widely separated taxa[3], added to which there is a tendency (understandable in the absence of any other clue) to look for familiar functions in preference to exploring novel paths. There are problems here, both intellectual and methodological, which are fundamental to phylogenetic analysis. Insulin, it should be added, is only one amongst many immunoreactive materials resembling vertebrate-type peptides which have been identified in invertebrates, but one other example must serve. The ascidian (protochordate) *Styela clava* is believed, on immunological and experimental evidence, to secrete a peptide with CCK-like properties, associated with a receptor system more generalised than that of the vertebrate alimentary tract, but capable of recognising vertebrate peptides[50].

Nor is such evidence confined to the chordate line. CCK is not easily distinguishable immunologically from gastrin, another alimentary hormone with which it shares the same pentapeptide sequence, but CCK/gastrin-like material has been demonstrated immunocytochemically in invertebrates as diverse as the ectoproctan *Bugula,* an earthworm, hydra, and an anthozoan[29]. But the evidence of wide distribution of certain peptides extends also to unicellular organisms and prokaryotes. Radioimmunoassay and bioassay have identified insulin-like material in the protozoan *Tetrahymena* and in the prokaryote *Escherichia coli,* as well as in two fungi *(Neurospora* and *Aspergillus),* while corticotropin-like and somatostatin-like immunoreactions have also been found in *Tetrahymena*[30]. It remains to be shown whether or not such materials function in any form of chemical communication, but evidence already suggests that receptors and effector pathways which would be expected to be associated with them are also present[43].

Challenging perspectives

This line of thought implies that the fundamental biochemistry of chemical communication must be very ancient indeed, and that phylogenetic advances have been founded upon an initially limited range of messenger molecules. The initial selection of these would have demanded a capacity to associate with receptors which can be visualised as having perhaps evolved out of protein subunits already present in cell membranes, and capable of assembly into appropriate configurations[11]. There is evidence that such associations already exist in unicellular organisms such as *Tetrahymena.* Phylogenetic exploitation of these associations and of their constituent molecules in multicellular forms would have involved, as already suggested, the functional diversification of endocrine and nervous systems out of a common ancestral complex[43], in which pluripotent cells of epithelial origin provided for external relations as well as for internal regulation[48]. Of course, there is still much that is speculative in these arguments. But they are opening up challenging perspectives, and they demand testing through a much wider range of observations and experiments, drawing, let us hope, upon many more species than the few that have so far been used.

1 Acher, R., Evolution of neuropeptides. Trends Neurosci. *4* (1981) 226–230.
2 Barrington, E.J.W., Evolution of hormones, in: Biochemical Evolution and the Origin of Life. pp. 174–190. Ed E. Schoffeniels. Elsevier/North Holland, Amsterdam and London 1971.
3 Barrington, E.J.W., Evolutionary aspects of hormonal structure and function, in: Comparative Endocrinology, pp. 381–396. Eds P.J. Gaillard and H.H. Boer. Elsevier/North Holland, Amsterdam 1978.
4 Barrington, E.J.W., Hormones and evolution: After 15 years, in: Hormones, Adaptation and Evolution, pp. 3–13. Eds S. Ishii, T. Hirano and M. Wada. Jap. Sci. Soc. Press, Tokyo 1980.
5 Barrington, E.J.W., and Thorpe, A., The identification of monoiodotyrosine, diiodotyrosine and thyroxine in extracts of the endostyle of the ascidian *Ciona intestinalis.* Proc. R. Soc. B *171* (1965) 136–149.

6 Bentley, P. J., and Scott, W. N., The actions of aldosterone, in: General, Comparative and Clinical Endocrinology of the Adrenal Cortex, pp. 418–564. Eds I. Chester Jones and I. W. Henderson. Academic Press, London 1978.

7 Bern, H. A., Loretz, C. A., and Bisbee, C. A., Prolactin and transport in fishes and mammals. Prog. reprod. Biol. *6* (1980) 166–171.

8 Burns, B. G., Sangalang, G. B., Freeman, H. C., and McMenemy, H., Isolation and identification of testosterone from the serum and testes of the American lobster *Homarus americanus*. Gen. comp. Endocr. *54* (1984) 429–432.

9 Chauvet, J., Hurpet, D., Colne, T., Michel, G., Chauvet, M. T., and Acher, R., Neurohypophysial hormones as evolutionary tracers: Identification of oxytocin, lysine vasopressin and arginine vasopressin in two South American opossums *(Didelphis marsupialis* and *Philander opossum)*. Gen. comp. Endocr. *57* (1985) 320–328.

10 Clarke, K. U., Visceral anatomy and arthropod phylogeny, in: Arthropod Phylogeny, pp. 467–549. Ed. A. P. Gupta. Van Nostrand and Reinhold Co., London 1979.

11 Csaba, G., The present state in the phylogeny and ontogeny of hormone receptors. Horm. metab. Res. *16* (1984) 329–335.

12 De Clerck, D., Eechaute, W., Leusen, L., Dederick, H., and De Loof, A., Identification of testosterone and progesterone in haemolymph of larvae of the fleshfly *Sarcophaga bulleto*. Gen. comp. Endocr. *52* (1983) 368–378.

13 Dunn, A. D., Studies on iodoproteins and thyroid hormones in ascidians. Properties of an iodinating enzyme in the ascidian endostyle. Gen. comp. Endocr. *40* (1980) 484–493.

14 Duve, H., Thorpe, A., and Strausfeld, N. J., Cobalt-immunocytochemical identification of peptidergic neurons in *Calliphora* innervating central and peripheral targets. J. Neurocytol. *12* (1983) 847–861.

15 El-Salhy, M., Falkmer, S., Kramer, K. J., and Speirs, R. D., Immunocytochemical evidence for the occurrence of insulin in the frontal ganglion of a Lepidopteran insect, the Tobacco Hornworm Moth *(Manduca sexta* L.) Gen. comp. Endocr. *54* (1984) 84–88.

16 Etkin, W., and Kim, Y. S., Role of the thyroid in vertebrate evolution, in: Evolution of Vertebrate Endocrine System, pp. 233–246. Eds K. T. Pang and A. Epple. Texas Tech. Press, Lubbock TX.

17 Frieden, E., The dual role of thyroid hormones in vertebrate development and calorigenesis, in: Metamorphosis, 2nd edn, pp. 545–563. Eds L. I. Gilbert and E. Frieden. Plenum Press, New York 1981.

18 Fritsch, H. A. R., Van Noorden, S., and Pearse, A. G. E., Gastrointestinal and neurohormonal peptides in the alimentary tract and cerebral complex of *Ciona intestinalis* (Ascidiaceae). Cell Tissue Res. *223* (1982) 369–402.

19 Grimm-Jørgensen, Y., Immunoreactive somatostatin in two pulmonate gastropods. Gen. comp. Endocr. *49* (1983) 108–114.

20 Grimm-Jørgensen, Y., Connolly, S. M., and Visser, T. J., Effect of thyrotropin-releasing hormone and its metabolites on the secretion of sulphated polysaccharides by foot integument in a pond snail. Gen. comp. Endocr. *55* (1984) 410–417.

21 Hartley, B. S., Evolution of enzyme structure. Proc. R. Soc. B *205* (1979) 443-452.

22 Heffmann, E., Functions of steroids in plants. Prog. Phytochem. *4* (1977) 257–276.

23 Hildebrand, J. G. C., Chemical signalling in the insect nervous system, in: Ciba Foundation Symp., vol. 88, pp. 5–11.

24 Iversen, L. L., Amino acids and peptides: fast and slow chemical signals in the nervous system? Proc. R. Soc. B. *221* (1984) 245–260.

25 Kanazawa, A., and Teshima, S., *In vivo* conversion of cholesterol to steroid hormones in the spiny lobster *Panulirus japonica*. Bull. Japan. Soc. Sci. Fish. *37* (1971) 891–898.

26 Kawauchi, H., Kawazoe, L., Adachi, Y., Buckley, D. I., and Ramachandran, J., Chemical and biological characterization of salmon melanocyte-stimulating hormones. Gen. comp. Endocr. *53* (1984) 37–48.

27 Kobayashi, H., Tsuneki, K., Akiyoshi, H., Kobayashi, Y., Nozuki, M., and Ouji, M., Histochemical distribution of peroxidase in ascidians, with special reference to the endostyle and the branchial sac. Gen. comp. Endocr. *50* (1983) 172–182.

28 Lance, V., Hamilton, J. W., Rouse, J. B., Kimmel, J. R., and Pollock, H. G., Isolation and characterization of reptilian insulin, glucagon and pancreatic polypeptide: complete amino acid sequence of alligator *(Alligator mississippiensis)* insulin and pancreatic polypeptide. Gen. comp. Endocr. *55* (1984) 112–124.

29 Larson, B. H., and Vigan, S. R., Species and tissue distribution of cholecystokinin/gastrin-like substances in some invertebrates. Gen. comp. Endocr. *50* (1983) 469–475.

148

30 LeRoith, D., Shiloach, J., Roth, J., and Lesnick, M. A., Evolutionary origins of vertebrate hormones. Proc. natn. Acad. Sci. USA 77 (1980) 6184–6188.

31 Licht, P., Evolutionary divergence in the structure and function of pituitary gonadotropins of tetrapod vertebrates. Am. Zool. 23 (1983) 673–683.

32 Martin, G., Keller, R., Kegel, G., Besse, G., and Jaros, P. P., The hyperglycaemic neuropeptide of the terrestrial isopod, Porcellio dilatatus. I. Isolation and characterization. Gen. comp. Endocr. 55 (1984) 208–216.

33 Mayr, E., The triumph of evolutionary synthesis. Times Literary Supplement 2 November (1984) 1261–1262.

34 Mordue, W., and Stone, J. N., Insect hormones, in: Hormones and Evolution, pp. 215–271. Ed. E. J. W. Barrington. Academic Press, London 1979.

35 Muggeo, M., Ginsberg, B. H., Roth, J., Neville, D. M., De Meyts, P., and Kohn, C. R., The insulin receptor in vertebrates is functionally more conserved during evolution than insulin itself. Endocrinology 104 (1979) 1393–1402.

36 Nagasawa, H., Kataoka, H., Isogai, A., Tamura, S., Suzuki, A., Ishizaki, H., Mizoguchi, A., Fujiwara, Y., and Suzuki, A., Amino-terminal amino acid sequence of the silkworm prothoracotropic hormone: homology with insulin. Science 226 (1984) 1344-1355.

37 Nakanischi, S., Inoue, A., Kita, T., Nakanura, M., Ohant, A. C., Cohen, F. M., and Numa, S., Nucleotide sequence of cloned cDNA for bovine corticotropin-β-lipotropin precursor. Nature, Lond. 278 (1979) 423–427.

38 Nozuki, M., and Gorbman, A., Distribution of immunoreactive sites for several components of pro-opiocortin in the pituitary and brain of adult lampreys, Petromyzon marinus and Entosphenus tridentatus. Gen. comp. Endocr. 53 (1984) 335–352.

39 Obermüller-Wilen, H., A neurosecretory system in the brain of the lancelet. Acta zool. (Stpckh) 60 (1979) 187–196.

40 Obermüller-Wilen, H., and van Veen, T., Monoamines in the brain of the lancelet Branchiostoma lanceolatum (Cephalochordata). Cell Tissue Res. 221 (1981) 245–256.

41 Plisetskaya, E., Kazakov, V. V., Soltitskaya, L., and Leibson, L. G., Insulin-producing cells in the gut of freshwater bivalve molluscs Anodonta cygnea and Unio pictorum and the role of insulin in the regulation of their carbohydrate metabolism. Gen. comp. Endocr. 35 (1978) 133–145.

42 Reinecke, M., Immunohistochemical localization of polypeptide hormones in endocrine cells of the digestive tract of Branchiostoma lanceolatum. Cell Tissue Res. 219 (1981) 445–456.

43 Roth, J., LeRoith, D., Shiloach, J., Rosenzweig, J. L., Lesnick, L., and Havrankova, J., The evolutionary origins of hormones, neurotransmitters, and other extracellular chemical messengers. New Engl. J. Med. 306 (1982) 523–527.

44 Sanders, B., Insulin-like peptides in the lobster Homarus americanus. III No glucostatic role. Gen. comp. Endocr. 50 (1983) 378–382.

45 Sandor, T., and Mehdi, A. A., Steroids and evolution, in: Hormones and Evolution, pp. 1–72. Ed. E. J. W. Barrington. Academic Press, London 1979.

46 Sawyer, W. H., Deyrup-Olsen, I., and Martin, A. A., Immunological and Biological characteristics of the vasotocin-like activity in the head ganglion molluscs. Gen. comp. Endocr. 54 (1984) 97–108.

47 Scarborough, R. M., Jamieson, G. C., Kalish, F., Kramer, S. J., McEnroe, G. A., Miller, C. A., and Schooley, D. A., Isolation and primary structure of two peptides with cardioacceleratory and hyperglycaemic activity from the corpora cardiaca of Periplaneta americana. Proc. natn. Acad. Sci. USA 81 (1984) 5575–5579.

48 Scharrer, B., Peptidergic neurons: facts and trends. Gen. comp. Endocr. 34 (1978) 50–62.

49 Thorndyke, M., Evidence for a mammalian thyroglobulin in the endostyle of the ascidian Styela clava. Nature, Lond. 271 (1978) 61–62.

50 Thorndyke, M., and Bevis, PJR., Comparative studies on the effects of cholecystokinin, caerulein, bombesin 6–14 nonapeptide, and physalaemin on gastric secretion in the ascidian, Styela clava. Gen. comp. Endocr. 55 (1984) 251–259.

51 Van Noorden, S., and Polak, J. M., Hormones of the alimentary tract, in: Hormones and Evolution, 2, pp. 791–828. Ed. E. J. W. Barrington. Academic Press, London 1979.

52 Wallis, M., The molecular evolution of pituitary hormones. Biol. Rev. 50 (1975) 35–98.

53 Wright, G. M., Immunocytochemical study of growth hormone, prolactin, and thyroid-stimulating hormone in the adenohypophysis of the sea lamprey, Petromyzon marinus L., during its upstream migration. Gen. comp. Endocr. 55 (1984) 269–274.

A new approach to the molecular evolution of hormones: the receptorial aspect

G. Csaba

Department of Biology, Semmelweis University of Medicine, POB 370, H–1445 Budapest (Hungary)

Studies into still existing organisms have suggested that the receptors and the hormones have themselves been subject to a process of evolution. The hormones of higher organisms, including humans, are only partly identical with those of lower organisms, but are nevertheless similar to these, and the existing dissimilarities seem to be the results of certain modifications[1,2]. Hormone formation in the higher organisms sets an example in how the combination or transformation of certain molecules, acting in themselves as hormones at the lower phylogenetic levels, gives rise to higher hormones, as is the case with thyroxine and its precursors. Since certain molecules which represent hormones in higher organisms do not function as such in lower ones, the hypothesis lies close at hand that these molecules had formerly been responsible for some other functions, which may have been non-hormonal for lack of an adequate receptor. Nevertheless, molecules capable of signal reception had necessarily existed already at the unicellular level[3], at which the range of the signals had in all probability been much wider, partly for lack of specialized signal receiver structures, partly for the necessity of recognizing all environmental molecules as useful or noxious, etc.[4–6].

The signal molecule potential of amino acids

Most hormones have arisen from amino acids, either simply by transformation (amino acid hormones), or by a reassembly of these (polypeptide hormones). Other hormones arose by linkage of certain active groups to the sterane structure. Since the amino acids represent fundamental nutriments or protein components for all living beings, the existence of amino acid receptors may well be taken into consideration at the lowest levels of phylogenesis. According to Lenhoff[15, 16], the receptors for amino acid hormones may have developed from the primitive amino acid (nutrient) receptors, in the course of the transformation of certain amino acids to hormones. Lenhoff based this hypothesis on his experiments with the hydra, but the same interrelationship has later been substantiated in experiments on the unicellular *Tetrahymena*, both by direct, and by imprinting-like effects. The latter are of particular importance since, in principle, a molecule capable of inducing imprinting is also capable of transforming to a signal molecule (hormone), because the cell 'remembers' the primary interaction with it, forms or transforms receptors for it, and alters thereby its response to the potential signal molecule.

150

Histamine functions in the higher organisms as a phagocytosis stimulating hormone[13, 14]. Its basic amino acid, histidine, stimulates the phagocytosis of the *Tetrahymena* both qualitatively and quantitatively to the same extent as histamine itself (fig. 1). However, preexposure of the unicellular to histidine inhibits completely the phagocytosis stimulating effect of subsequent histamine exposure, from which it follows that the two active molecules bind in all probability, to the same receptor[7]. Amplification with histidine imprints the receptor for histidine reception, which results in irresponsiveness to histamine. The effect of histidine is at the same time more universal, for it stimulates not only the phagocytic activity, but also the division of the protozoon, on which histamine has no effect whatever (fig. 2). It appears that transformation of a potential signal molecule to a hormone involves an increase in the specificity, but simultaneously also a decrease in the range, of action. The fact that preexposure to histamine inhibits later interaction with histidine exactly as the latter inhibits

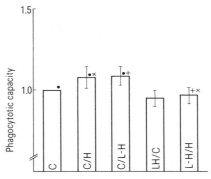

Figure 1. Effect of histamine (H) and histidine (L−H) on the phagocytotic capacity of *Tetrahymena*. Significance: . = C vs C/H; C vs C/L−H (p < 0.05), + = C/L−H vs L−H (p < 0.02), x = C/H vs L−H/H (p < 0.05).

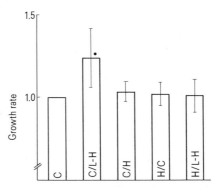

Figure 2. Effect of histamine (H) and histidine (L−H) on the growth rate of *Tetrahymena*. Significance: . = C vs C/L−H (p < 0.05).

interaction with the former, presents an additional proof that the two related molecules bind to the same receptor. This weighs in favor of the hypothesis that the histamine receptor developed from a primitive amino acid (nutrient) receptor. Similar phenomena have been observed with tyrosine and diiodotyrosine. The hormones (precursors) of the tyrosin series stimulate cell division, also at the level of the *Tetrahymena,* for which diiodotyrosine represents the most active mitosis stimulating hormone of the series[9], but diiodotyrosine has a similar effect, and the two molecules inhibit each other's action on subsequent exposure exactly as do histidine and histamine.

The foregoing considerations seem to support the implication that not only hormones, but also amino acids could induce imprinting, which would result in an increased response of the cell to reexposures. However, experimental facts have disproved the validity of this conclusion for all amino acids. Since the imprinting potential seems to depend greatly on the capability of transformation to a signal molecule, investigations into the imprinting potential of certain amino acids – at the lower levels of phylogenesis – can probably throw more light on the problem.

In the higher organisms, genuine hormones have arisen from tyrosine, phenylalanine (triiodothyronine, thyroxine, epinephrine, norepinephrine) tryptophane and histidine (the tissue hormones serotonin and histamine), whereas glycine and some other molecules are functioning as neurotransmitters. Examination of these molecules, and of the polypeptide hormone components serine and valine, which have no hormone-like action in themselves, for imprinting potential, has substantiated a genuine positive imprinting effect exclusively in the case of serine, and a negative imprinting in the case of glycine[8]. Of the aromatic amino acids studied, tyrosine and phenylalanine stimulated the growth of the *Tetrahymena,* whereas the indole ring containing heterocyclic amino acid tryptophane had no influence on it (fig. 3). This suggests the

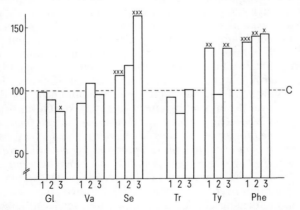

Figure 3. Effect of various amino acids on the growth of *Tetrahymena* (related to the control as 100), in the case of the first (C/A) and second (A/A) encounter and 48 h after the first encounter (A/C). Gl, glycine; Va, valine; Se, serine; Tr, tryptophane; Ty, tyrosine; Phe, phenylalaline. 1) C/A, 2) A/C, 3) A/A. Significance: x, 0.05 > p > 0.02; xx, 0.02 > p > 0.01; xxx, p < 0.01.

superiority of the benzol ring containing aromatic amino acids to the indole ring containing one in influencing the division of the *Tetrahymena*. The molecules capable of inducing (positive or negative) imprinting, i. e. serine and glycine, are chemically similar small polar molecules, whereas the thoroughly ineffective molecule valine carries a polar side chain. Serine possesses a free alcoholic hydroxyl group, whereas glycine, alone of the molecules tested, is optically inactive. Serine is a very frequent component of proteins (and of polypeptide hormones as well), and glycine acts as a CNS-inhibitor (interestingly, its imprinting effect also is negative).

The above experimental observations seem to support the hypothetical implication that the receptors of amino acid hormones may arise among others from nutriment receptors. Nevertheless, the 'transformation to hormone' potential of the amino acids may vary with their chemical structure, capability of forming specific receptors for themselves, and with their suitability for inducing imprinting as well. As already noted, not all amino acids can give rise to imprinting, i. e. to a lasting stimulation of cell division not all of them stimulate the mitotic function (transmission of the information to a great number of progeny generations), on primary interaction, while others develop a powerful action to this end.

The signal molecule potential of oligopeptides

Exactly as relatively few amino acids developed to amino acid hormones or neurotransmitters (or tissue hormones), the known 'families' of polypeptide hormones also are moderate. This can be explained either by development of a family from a single molecule by modification (e. g. mutation), or by a greater suitability of certain molecule types (amino acid sequences) for the signal molecule function, compared to others. Probably certain structures (amino acid sequences or the configurations induced by these) induce receptor formation (presumably through the imprinting mechanism) more efficiently than others with a reduced or even lacking potential to this effect.

In experimental conditions, dipeptides proved to be as efficient inducers of imprinting (i. e. of receptor formation or amplification) as certain amino acids, with the additional advantage that increase of the chain length and, especially, coupling with a terminal tyrosine group, increased the efficiency of imprinting enormously[10]. On the other hand, interchange of amino acids inside the chain also had influence on the imprinting effect. In any event, *the molecule inducing the most powerful imprinting in the Tetrahymena and in cell lines as well, was always the one noted for strongest physiological effect. It follows that, from the point of view of hormone evolution, the receptor forming potential of the molecule or sequence plays the key role in transformation to a signal molecule.*

However, the amino acid components of the oligopeptide also play an important role. The terminal tyrosine group seems to develop an extraordinarily powerful receptor forming effect, and proline, too, seems to have a decisive influence. This secondary amine, carrying its N-atom in a rigid ring structure, is involved in forming the steric structure of proteins by inhibiting the stabilization of the normal alpha helix, whose structure disintegrates in presence of the proline residue[2]. Although direct evidence is lacking, these properties could well account for a powerful imprinting action as well.

It appears that those peptides, whose specific and nonspecific binding functions are well segregated, i.e. which show little tendency, if any, to cross-reactions, have the strongest signal molecule potential[11].

The ontogenetic development of higher organism naturally takes place in presence of predetermined receptor structures and hormones, but this does not exclude the possibility of changes in the amino acid sequence of a given hormone, or of the slightly dissimilar amino acid sequence of a related hormone, for variations in the intensity of imprinting. Moreover, taking into consideration species-specific differences in certain amino acid sequences, a perinatal hormone treatment (e.g. pig insulin for the rat chicken or human beings) may also alter the trend or the intensity of imprinting. Accordingly, investigations into the imprinting effect of amino acids and amino acid hormones, or of various synthetic oligopeptides of predetermined structure, may simultaneously throw a light on the molecular evolution of hormones, and on the perinatal impact of hormone variants.

1 Barrington, E. J. W., Introduction, in: Hormones and Evolution, vol. 1, pp. VII–XXI. Ed. E. J. W. Barrington. Academic Press, London/New York 1979.
2 Conn, E. E., and Stumpf, P. K., Outlines of Biochemistry, pp. 75 and 94. Wiley and Sons, New York/Toronto 1976.
3 Csaba, G., Phylogeny and ontogeny of hormone receptors: the selection theory of receptor formation and hormonal imprinting. Biol. Rev. 55 (1980) 47–63.
4 Csaba, G., Ontogeny and Phylogeny of Hormone Receptors. Karger, Basel/New York 1981.
5 Csaba, G., The unicellular Tetrahymena as a model cell for receptor research. Int. Rev. Cytol. 95 (1985) 327–377.
6 Csaba, G., Why do hormone receptors arise? in: Development of Hormone Receptors. Ed. G. Csaba. Birkhäuser Verlag, Basel 1987.
7 Csaba, G., and Darvas, Zs., Receptor-level interrelationships of amino acids and the adequate amino acid type hormones in Tetrahymena: a receptor evolution model. BioSystems 19 55–59.
8 Csaba, G., and Darvas, Zs., Hormone evolution studies: multiplication promoting and imprinting ('memory') effect of various amino acids. BioSystems, in press (1987).
9 Csaba, G., and Németh, G., Effect of hormones and their precursors on protozoa: the selective responsiveness of Tetrahymena. Comp. Biochem. Physiol. 65B (1980) 387–390.
10 Csaba, G., Kovács, P., Török., O., Bohdaneczky, E., and Bajusz, S., Suitability of oligopeptides for induction of hormonal imprinting in a cell line and in Tetrahymena. Implication on receptor and hormone evolution. BioSystems 19 (1986) 285–288.
11 Csaba, G., Kovács, P., Tóth, S., and Bajusz, S., Overlapping imprinting of oligopeptides in Chang liver cells. Data on the mechanism of hormone evolution. BioSystems, in press (1987).
12 Gorbman, A., Dickhoff, W. W., Vigna, S. R., Clark, N. B., and Ralph, C. L., Comparative Endocrinology, p 33. Wiley and Sons, New York/Toronto 1983.
13 Jancsó, M., Speicherung. Publ. House Hung. Acad. Sci., Budapest 1955.
14 Kushinsky, G., Hille, W., and Shimassek, A., Über Histamin als Mittersubstanz bei der Wirkung von Adrenochrom als Aktivator von Reticuloendothelzellen. Arch. Pharmak. 266 (1955) 1–8.
15 Lenhoff, H. M., Behavior, hormones and hydra. Science 161 (1968) 434–442.
16 Lenhoff, H. M., On the mechanism of action and evolution of receptors associated with feeding and digestion, in: Coelenterate Biology, pp. 238–239. Eds L. Muscatine and H. M. Lenhoff. Academic Press, New York 1974.
17 Lison, L., and Smoulders, J., Action de l'histamine sur le système réticuloendothelial chez la grenouille. C. r. Séanc. Soc. Biol. 143 (1949) 575–577.

Common origin and phylogenetic diversification of animal hormonal systems

D. Bückmann

Abteilung Allgemeine Zoologie, Universität Ulm, D–79 Ulm/Donau (Federal Republic of Germany)

The significance of the evolutionary history of hormones

The evolutionary history of hormones can contribute to our understanding of hormonal function. How does a compound become a hormone?

The known animal hormones are members of different molecular groups, such as amines, peptides, steroids, juvenoids and iodinated thyronines. However, none of these groups consists only of hormones. There are amines, peptides, and steroids without hormonal properties. Evidently belonging to a certain chemical class of compounds is not sufficient for a molecule to be a hormone.

To enable a compound to be a hormone a whole functional system is required, consisting of the compound itself, the hormone-producing tissue, the target organ, the primary effect of the hormone receptor complex and the final biological function. These factors together have been called the components of a 'hormonal system' by Karlson and Gersch[12,33].

There is a great variety of hormonal systems among metazoan animals, including man. The organs producing hormones and reacting to hormones have evolved from diverse other structures, which originally may have had other functions. They must have evolved in a way in which they always remained functional. Otherwise the organisms would have died out, and none of their progeny would exist today; only such progeny can be investigated physiologically. Therefore some properties of hormonal systems may just have historical causes, which cannot be understood without knowing their evolutionary history. An example for this is the history of the thyroid and its iodine-containing hormones as revealed by Gorbman[27] and Barrington[3].

The evolutionary history of hormonal systems may be reconstructed by comparing the hormones of animals with various degrees of relationship, and tracing back the phylogenetic branches of animals with different hormones to common ancestors. In this way one can find out at what branching-point of the phylogenetic tree a change must have occurred. From what is known about animal evolution from fossils, one can even tell how many millions of years ago such an event took place. An impressive example for such a reconstruction is the history of the neurohypophyseal hormones as revealed by Acher[1].

In order to clarify the history of hormonal systems one could try to make a comparative survey of the hormonal organs, substances, and functional systems. All these components must have coevolved in such a way that the whole system remained functional. Therefore one might expect that a comparative study of the hormones, the hormonal organs, the target organs and the biological functions might all point to phylogenetical systems identical to those of the animals.

Surprisingly, this was not the case, when whole animal classes and phyla were compared[11,12]. There were rather certain rules as to the extent of diversity between the glands, the chemical compounds, and the functions in different animal groups. From these rules conclusions could be drawn on the phylogenetic age of the differences between those components, using the method outlined above.

This method led to the conclusion that the peripheric epithelial hormonal glands releasing nonpeptide hormones must have evolved late in metazoan evolution, after the branching-off of the main animal phyla, and independently in each of them[12,51].

This result was challenged immediately[12]. It was not so attractive as an explanation involving one general developmental step common to all animals would have been. A monocausal hypothesis explaining every development by the same general principle is always the most attractive. Equally attractive is the picture of the evolution of physiological mechanisms from primitive ancestors continuously reaching higher levels and arriving at their most evolved state in the human species.

This picture has severely hampered progress in comparative endocrinology. It did not allow invertebrates to possess other hormonal systems than vertebrates. When, in insects, the search for sex hormones like those in mammals remained without success, it was generally believed that invertebrates do not possess any hormones at all, though already in 1917 and 1922 Kopec[39,40] had claimed that the insect brain might have a similar function to the vertebrate thyroid in promoting metamorphosis. In fact this was the first example of neurosecretion. However, he could not prove his results sufficiently and there was no general scientific concept into which they would have fitted. So his work was not taken notice of.

The approach of comparative physiology

A new concept was provided by the idea that different animal phyla are not so much different in their level of perfection, but that they are constructed according to different principles. It is characteristic that the first book on comparative physiology, by von Buddenbrock 1928[7a] revived the discussion about insect hormones. From the concept mentioned von Buddenbrock concluded that hormones of other animal phyla might have functions different from those of vertebrate hormones[14].

Hormones reach every body cell equally well and equally fast through the blood. They are well-fitted for synchronizing events in different and distantly located tissues. In order to find invertebrate hormones one had to look for such a synchronization. In vertebrates synchronization of sexual characters is necessary due to their periodical phases of reproduction, whereas in insects all the adult life is solely dedicated to reproduction. Therefore it is quite sufficient when the genetic mechanism of x and y chromosomes present in every body cell causes either the male or the female form to arise, lasting for the whole adult life-span. What seemed likely to require synchronization in arthropods was molting and color change[14].

Not knowing the source of the hormones which might control these events, von Buddenbrock chose the method of blood injection. His student Gottfried Koller adduced the first real proof for invertebrate hormones by injection experiments on the color change of shrimps[37,38]. For molting hormones the injection method was not so successful, but it brought new indications that they existed[7-9] and started the general discussion and investigation of those hormones[55].

Thus for our understanding of the evolution and diversification of hormonal systems in metazoan phylogenetic history the comparison of hormonal functions seems to be more fruitful than that of glands.

The common base of intercellular messenger substances in multicellular organisms

The new results and concepts of the evolution of receptor systems compiled by Csaba[18a,b] and Bradshaw and Gill[6] present a new basis for consideration on the evolution of hormonal systems.

They show that the receptor mechanism is a property of all cell membranes, common to procaryotes as well as to eucaryotes[15,17,18,31]. Equally important is the finding that new receptors for external compounds can be developed every time by imprinting[18,41]. Furthermore, an increasing number of results show that compounds which serve as hormones in certain mammalian tissues can be found in other tissues too[26], in other animal phyla[2], and, as has been investigated extensively by the group of Roth and Le Roith[43a,b], even in Protoza. Mammalian peptide hormones, as well as receptors for them[20], are also present in plant tissues[17,50,54].

All this means that the different components of hormonal systems are not of the same phylogenetic age. The principle of receptor mechanism, the ability to develop receptors anew, and even many compounds suitable to be hormones were already in existence at the onset of metazoan evolution.

All these components were prerequisites for the development not only of hormones, but of intercellular messenger substances in multicellular organisms in general, such as chalones, prostaglandins, morphogenetic substances, poietins and tissue-specific growth factors. In fact, in investigations on their receptor mechanism, these growth factors are no longer distinguished from hormones[6].

However, evidence is accumulating that there may be a basic difference between the receptor mechanisms for peptides[15] and those for steroid hormones, juvenoids and thyroid hormones, as the latter ones may be restricted to the inside of the nucleus only[28], while the former are not found in the nucleus.

It has been proposed to confine the term 'hormone' to messenger substances acting within the nucleus on gene activation only. However, this might include substances of monocellular organisms, whereas hormones are intercellular messengers in multicellular bodies. On the other hand, it would exclude many substances presently known as hormones. Such a definition would cause confusion. A definition referring to features which really came into existence during the evolution of hormones in multicellular organisms might be preferred.

The different evolutions of animal and plant hormonal systems

Evolution of differentiated multicellular organisms has occurred twice, in plants and animals.

The organization of higher plants makes conditions for release, transport and functioning of hormones quite different from those in multicellular animals. Special features are the air-filled intercellular spaces and the cell walls which prohibit direct contact of cell membranes while the cells are connected by plasmodesms. Indeed, there are some indications that phytohormones have had an independent phylogenetic development different from that in animals.

While higher plants possess all the chemical groups and even the individual substances, like peptides and steroids, which serve as hormones in animals, they do not make use of them as hormones. Instead they have evolved their own phytohormones like gibberellins, indolyl acetic acid and kinetins. Possibly most of them are ubiquitous plant metabolites which subsequently have become hormones by the development of receptors for them[49].

For all these reasons, the question remains open how far plant hormone receptors are different from those in animals. There seem to be difficulties in isolating them with the same methods as are used for animal receptors.

Another difference is in the dose response curves. Since animal hormones have to be bound to receptors in order to act, their dose response curve is a saturation curve[42]. Saturation is achieved when all receptors are occupied, physiological concentrations ranging from about 10^{-9} to 10^{-7} M. However, dose response curves of plant hormones are reported to increase continuously up to 10^{-4} M, eventually turning down again as an optimum curve[34].

A special question is that of fungal hormones, because some pheromones of primitive fungi are related to vertebrate steroids[36] and possibly there is, indeed, a phylogenetic relationship between animals and fungi[23].

Diversification of hormonal systems in multicellular animals

While there are possibly differences between plant and animal receptor mechanisms, this seems not to be the case among the receptors of various animal groups. However, not much information is yet available about the isolation and the properties of invertebrate hormone receptors.

As receptor mechanisms seem to be essentially similar for all hormones, they can help us to understand that, in principle, a further evolution of hormones was possible in metazoan phyla[4], but they cannot help us to find out how the present differences among animal phyla originated.

There is, indeed, a common root of metazoan hormonal systems. This is the existence of the receptor system, the imprinting mechanism, and the substances required. What occurred during metazoan evolution of hormonal systems was their functional connection. But, as the components mentioned were present all the time, hormonal systems could arise at different phylogenetic stages. Thus we can expect hormonal systems of different phylogenetic ages to exist.

The concept of the comparative method is the assumption that structures common to all metazoans were probably evolved in ancestors common to all of them, which means at the beginning of metazoan evolution, while structures which are peculiar to only one phylum or class probably evolved only in its own ancestors after the branching off from the phylogenetic tree.

From the comparative method we can expect an answer to the question as to when hormonal systems arose in phylogenetic history. From properties common to different animal groups possessing similar hormonal systems we can also draw some conclusions about under what conditions hormonal systems arise and why.

The evolutionary history of neurosecretion

Common to all metazoan phyla is the existence of neurosecretion. Substances with the properties of neurosecretion can be shown even in sponges, parazoa, which do not possess real typical neurons[44]. Thus, neurosecretion seems to be phylogenetically older than the typical properties of neural cells with axons, fast conduction of membrane depolarization and release of transmitter substances. We can imagine how this developed. At first there was a coordination by messenger substances, which had to be moved through the multicellular body over long distances. Therefore, they had to be long-lasting large molecules. They needed a very specific conformation which could be recognized only by those cells meant to react to them.

When increasing body size and level of differentiation asked for a quicker and more exacting mechanism, the cells responsible for coordination grew extensions towards the target cells, only contacting those cells which had to react. Thus the typical properties of neurons arose. Because of the presence of axons, long distance transport of messenger substance and the danger of being answered by the wrong cells were avoided. The transmitter substances could be small, short-lived, and unspecific.

However, in cases were many different cells at different positions had to be informed, the old transport of specific molecules retained its advantages. The cells producing such signal peptides even at the same time retained the capability of producing small amines like the transmitter substances by decarboxylation of amino acids, the so-called APUD characteristics (from Amino acid and Precursor Uptake and Decarboxylation), of Pearse[45].

These characteristics are, however, also found in other tissues than the nervous system. These tissues have acquired the ability of forming peptide hormones, too. As such sources of hormones seem not to be common to all phyla, they must be phylogenetically younger.

At the same time the nervous system itself improved its capability of hormonal coordination by a more exact control of hormone release into the blood. Neurohemal structures were evolved[52], like the neurohypophysis of vertebrates, the corpora cardiaca of insects, the sinus gland of decapod crustaceans, and certain structures in molluscs[32,53] and annelids[22,30].

The fact that such structures are not well developed in all metazoan phyla, again points to the conclusion, that they developed independently in the phyla mentioned. In this case there is still another indication of phylogenetic independence; their structures are extremely different in the different groups.

The vertebrate neurohypophysis and the crayfish sinus gland are rather compact and are in the neighborhood of large blood vessels. In insects Raabe discovered the diffuse peripheral endocrine system[47]. This may be connected with tracheal respiration which does not require a very effective blood circulation in insects. There are many small sites of hormone release distributed all over the insect peripheral nervous system. In molluscs large areas in the wall of the blood vessels of whole body regions serve as neurohemal structures. Moreover, in snails there are most complicated special neurosecretory elements such as the 'canopy cells'[32]. Such differences may have to do with the absence of a blood-brain barrier.

Neuroendocrine and epithelial endocrine glands

The phyla possessing highly developed neurohemal organs have, in addition, evolved neuroendocrine glands. These glands do not derive from neural tissue but are closely adjoining the main ganglia, like the vertebrate adenohypophysis, the insect corpora allata and the cephalopod optical glands.

Furthermore, these phyla are the only ones which comprise groups in which typical peripheral epithelial hormone glands secreting non-peptide hormones have been found, such as the known mammalian hormone glands and the molting glands of arthropods.

There are three groups of nonpeptide hormones, the steroids, the juvenoids, and the iodine-containing vertebrate thyroid hormones. Of these only steroids have so far been found serving as hormones in several different animal phyla, but not in all of them. However, the juvenoids are known as hormones only in one class of the arthopod phylum, the insects, and iodinated thyronines only in one subphylum of the chordates, the vertebrates.

These nonpeptide hormones are invariably released by peripheral epithelial endocrine glands. Thus again one has to conclude that, if there is a real borderline between different hormonal systems, it might be between peptide and steroid hormones.

Under what conditions do complicated hormonal systems evolve?

As functions of temporal coordination and synchronization occur in every multicellular organism, one may ask why these structures evolved only in some of them. Neurohemal organs and peripheral endocrine glands are not generally distributed, even within those phyla where they occur. They are always a property of highly differentiated and complicated groups like malacostracan crustaceans, insects, cephalopods, and gastropoda, but are missing in other classes of the same phyla like entomostracan crustaceans, simple arachnids, scaphopods and tunicates.

'A high level of organization or differentiation' is a highly subjective judgement. However, it can be replaced by an objective parameter, the body size. Since the linear dimensions increase in proportion to the body length, but the surfaces of the body and its organs increase with the square, and their volume (which means the mass) even with the cube of the linear dimension, a small change in body size causes considerable qualitative differences. Organisms cannot just be enlarged or made smaller like the people in the story of Gulliver. Large body size poses special problems for the organs of mechanical stabilization, of transport, and of coordination, among them the hormonal systems.

It is not just coincidence that the largest existing animals, the vertebrates, do not include really small, microscopical forms like all other animal phyla. Body size and level of organization are related, and a certain level of organization requires at least a certain body size. This corresponds to the highly complicated hormonal system controlling reproduction in vertebrates.

Hormonal control of sexual differentiation

Indeed, hormonal control of reproduction is a good example for diversification of hormonal systems in different phyla, especially as the basic task of producing and maturing male and female gametes seems to be rather uniform within the animal kingdom. However, even the distribution of different mechanisms of phenotypic or genetic sex determination, such as X and Y chromosomes, and haploidy vs. diploidy among animal taxa, has been called 'chaotic' by Harvey and Partridge[29].

Equally different are the hormonal mechanisms[30]. We can, however, ascribe certain steps of organization of this hormonal system to certain steps in the general level of organization and body size[13].

In every case there are two phases of development; one of growth, usually accompanied by a good ability for regeneration, and one of sexual reproduction. On this level only one hormone is required, controlling the transit from the first to the second phase. This may be a juvenile hormone, the cessation of which will cause the transition, or a gonadotropic hormone, the appearance of which will cause the same effect. The resulting reproductive organisms would be simultaneously bisexually functioning gynanders.

The ability for regeneration can be so perfect that X asexual reproduction by buds occurs. In this case an alternation of sexual and asexual generations occurs, a metagenesis, as in many Coelenterates and certain polychaete annelids, like *Autolytus prolifer*[22].

When the differentiation of sexual characters reaches such a level that male and female characters cannot be developed at the same time, a second hormone is afforded, which controls a change of sex. A consecutive gynander will result. The new hormone might be either an androgenous or a gynogenous one.

The same hormonal outfit will still be sufficient, when different sex characters can no longer be expressed in the same body at all. Again, one sex is caused to develop by a hormone, while the other develops when this hormone is lacking. Examples for this are many malacostracan crustaceans with their androgenic hormone[16].

The most complicated system, with gonadotropic hormones and different sex hormones for the two sexes, seems to be restricted to the vertebrates only.

The evolution of endocrine organs from the target tissue

Another example for the relation of a hormonal system to the level of organization is presented by the molting glands of malacostracan crustaceans and the insect corpora allata, which control metamorphosis by their juvenoids. Both organs develop from parts of the integument. In stick insects the corpora allata can be seen to molt inwardly into their central cavity which represents the outer side of the integument[46], and the same can be seen in the crayfish molting gland[35]. This means that these glands are part of their own target tissue and respond to their own hormone. The same hormone is required in lower crustaceans and arachnid groups, in which no molting glands have been found so far. One can imagine that in these simpler forms the whole integument is capable of producing the ecdysteroid required for molting. In insects and higher crustaceans some of the cells have taken over this task and specialized in it. They produce the hormone for all the other cells and distribute it to them via the body fluid. This may mean a better and more exact triggering of the molt, which is afforded by the larger and more complicated exuvia of these groups[12].

The evolution from part of the target tissue, however, is not the only way of evolving an epithelial hormonal gland. Essential is the differentiation of a group of cells releasing a compound which is needed by others for a certain reaction. This compound has to be transportable by the body fluid, and recognition by receptors must be possible. These conditions seem to be fulfilled well by steroids, which are hormones in a number of different phyla. However, even rather unusual compounds can gain a hormonal function, like the iodine-containing vertebrate thyroid hormones. An idea about how they may have become hormones, based on comparative morphology and physiology, is presented by Barrington[3] and by Gorbman[27].

The principle of polytropic action

The concept of synchronization as the main task of hormonal systems even answers the question why comparative physiology has not succeeded in finding out which function of a hormone might be the 'original' one, which means the phylogenetically oldest and first function.

In the case of the molting gland and its hormone, the function seems to be rather consistent. It always serves as a molting hormone. However, the reactions evoked may be quite different. If a color change is required to hide a caterpillar while it is leaving the green food leaves in order to molt to a pupa, the molting hormone may at first, at a low concentration, act as color change hormone and later on, at a higher concentration, evoke the molt[10]. In dipteran flies the last but one larval cuticle becomes hard and brown before pupation as a shelter for the later pupa. In this case the hormone causes sclerotization. However, in some species the cuticle is hardened not by proteins but by calcium. In these species the hormone causes calcification[25]. The same hormone does quite the contrary in crayfish, were a decalcification of the old cuticle is the first step of molting. In some adult flies, which first have to move out of their substratum before their cuticle can harden, the molting hormone causes all the molting with the exception of the sclerotization process. This is, later on triggered by another hormone, bursicone[24]. Our knowledge about the explanation of this multiplicity has been called a 'tabula rasa' by Fraenkel[24].

Even more inconsistent are the metabolic and the metamorphosis effects of thyroid hormones.

What is common to all the effects of one hormone? It is the timing. Every process connected with a molt or with the ingestion of food has to be triggered at the same time. The synchronization is mainly afforded for different reactions in different tissues. If they were connected causally in another way, no hormone would be necessary.

Ecdysteroid causes molt and color change, and later on in adult development also gonadal maturing[19]: It has ecdysiotropic, melanotropic, and gonadotropic functions. It is truly polytropic[11]. The same applies to other hormones.

The vertebrate hormone prolactin causes quite different reactions of reproduction and osmotic responses in the different groups. Bern[5] was almost reproachful about this unprincipled 'versatility'. He assumed that the hormone may be a kind of emergency hormone for special stress situations. But the same might be said for other hormones, and the same explanation cannot be valid for all of them.

In many vertebrates reproduction is linked with changes in environment. Amphibians leave the dry land and enter the water. Many fishes migrate for reproduction from saltwater to freshwater or vice versa. Thus changes in osmoregulation really have to be synchronized with reproduction.

In his review on melatonin Epple[21] arrives at the conclusion that this hormone is always used when processes are controlled by diurnal rhythm. Again it is the timing which is common to one hormone's functions.

The concept of Barrington[3] and Gorbman[27] on the evolution of thyroid hormones is based on the homology of the endostyle of lower chordates with the vertebrate thyroid, which becomes directly apparent at the metamorphosis of the ammocoetes larva of lampreys, when the endostyle is transformed directly into the thyroid gland. In the lower chordates the endostyle, and also the whole body surface, produce mucus containing iodinated amino acids. As seawater is an environment rich in iodine this is common among many marine invertebrates. The endostyle slime is used for wrapping food particles. It is digested and resorbed by the alimentary tract together with the food. It can serve as a signal of the arrival of a food supply to the whole body. The fact that the function as a thyroid gland begins at metamorphosis also makes plausible the idea that its hormone can serve as a signal of metamorphosis, too.

It looks as though the timing of hormone release is even more consistently inherited in animal evolution that the actual effects.

With respect to cases like the caterpillar, which reacts at first to a small dose of a hormone by color change and, later on, to higher concentrations of the same hormone by pupation[10], instead of 'synchronization' the term of 'temporal coordination' might be more exact and appropriate.

The concept of temporal coordination would make it quite sensible to assume that hormones are, on principle, polytropic, because it is their very function to synchronize events as different to each other as possible. If this reasoning is correct, there should be no causal direct connection between the structure of a hormone and a certain mode of functioning. The hormone is merely suitable for transmitting information, just as a copper wire is, quite independently of the content of the telegrams transmitted.

The functional connection is established in a second step by the receptor mechanism. In different cases the hormone receptor complex can be coupled to different successive processes[12]. The hormone molecule has to be appropriate in that it occurs at a certain time, can be transported in the body fluid and can be recognized by receptors.

Summary

The comparative view leads to the following main conclusions:
1) Hormones are intercellular messengers in multicellular organisms. However, the receptor mechanism, the ability of forming receptors to substances in the cell's environment and the ability to synthesize most of the substances which serve as hormones in metazoans, are present in unicellular organisms, too.
2) The main achievement of multicellular organisms in evolving hormonal mechanisms is due to their ability of differentiation. Though the whole genome and the ability to synthesize certain substances is, in principle, common to all body cells, the forming of certain substances and the ability to react to them in a certain way is delegated to certain cell groups only. This may be common to many intercellular messenger substances such as chalons, prostaglandins, morphogenetic and tissue-specific growth substances. A special feature in hormonal systems is that the two sites of release and reaction are distinct and are located at a distance from each other. Possibly this is the main or even the only difference from other intercellular messengers. However, it is of great functional importance because it enables hormones to control the temporal coordination of entirely different processes in different tissues located at a distance from each other or distributed all over the body. This feature is common to all compounds presently known as hormones.
3) While the localization of receptors in target cells, as well as the nature of the releasing tissue or the mode of transport through the blood, may be too narrow borders for a definition (for instance in animal groups without a closed blood circulation system) the fact that there is transport over a distance is not. The fact that release and reaction sites are located at a distance from each other within the multicellular body may serve as a definition of 'hormones' based on a common phylogenetic root and functional importance.
4) On this common base different animal phyla have evolved different hormonal systems in relation to the particular problems of long distance temporal control of physiological processes posed by their special type of organization.

1 Acher, R., Molecular evolution of neurohypophyseal hormone and neurophysins, in: Neurosecretion and Neuroendocrine Activity. Evolution, Structure and Function. Eds W. Bargmann, A. Oksche, A. Polenow, and B. Scharrer. Proc. VII int. Symp. Neurosecretion. Springer, Heidelberg/Berlin/New York 1978.
2 Ball, I. N. (Ed.), Abstracts of the 13th Conference of European Comparative Endocrinologists. Belgrad, 3.–12. Sept. (1986) 1–191.
3 Barrington, E. J. W., Some endocrinological aspects of Protochordata, in: Comparative Endocrinology, pp. 250–265. Ed. A. Gorbman. Wiley, New York 1953.
4 Barrington, E. J. W., The phylogeny of the endocrine system. Experientia *42* (1986) 775–781.
5 Bern, H., On two possible primary activities of prolactins: osmoregulatory and developmental. Verh. dt. zool. Ges. *68* (1975) 40–46.

6   Bradshaw, R. A., and Gill, G. N. (Eds), Evolution of hormone-receptor systems. 11th Annual UCLA Symposia, Abstr. J. cell. Biochem. 6 (1982) 110–185.

7a v. Buddenbrock, W., Grundriss der vergleichenden Physiologie. Verlag Borntraeger, Berlin 1928.

7b v. Buddenbrock, W., Vergleichende Physiologie, vol. 4, Hormone. Birkhäuser Verlag, Basel 1950.

8   v. Buddenbrock, W., Beitrag zur Histologie und Physiologie der Raupenhäutung mit besonderer Berücksichtigung der Versonschen Drüsen. Z. Morph. Ökol. Tiere 18 (1930) 701–725.

9   v. Buddenbrock, W., Untersuchungen über die Häutungshormone der Schmetterlingsraupen II. Z. vergl. Physiol. 14 (1931) 415–428.

10 Bückmann, D., Die Auslösung der Umfärbung durch das Häutungshormon bei Cerura vinula (Lepidoptera, Notodontidae). J. Insect Physiol. 3 (1959) 159–189.

11 Bückmann, D., The Phylogeny and the Polytropy of Hormones. Horm. metab. 15 (1983) 211–217.

12 Bückmann, D., The Phylogeny of Hormones and Hormonal Systems. Nova Acta Leopoldina 56/255 (1984) 437–452.

13 Bückmann, D., Vergleichende Endokrinologie und Stammesgeschichte der Sexualentwicklung. Akt. Endokr. Stoffw. 5 (1984) 169–174.

14 Bückmann, D., Wolfgang von Buddenbrock und die Begründung der vergleichenden Physiologie. Medizinhistor. J. 20 (1985) 120–134.

15 Carpentier, J.-L., Gorden, P., Robert, A., and Orci, L., Internalization of polypeptide hormones and receptor recycling. Experientia 42/7 (1986) 734–744.

16 Charniaux-Cotton, H., Recouverte chez un Crustace amphipode (Orchestia gammarella) d'une glande endocrine responsable de la différentiation de caractères sexuels primaires et secondaires males. C.r. Acad. Sci. Paris 239 (1954) 780–782.

17 Chadwick, C. M., and Garrod, D. R. (Eds), Hormones, receptors, and cellular interactions in plants. Cambridge University Press, Cambridge, London, New York 1986.

18a Csaba, G., Why do hormone receptors arise? Experientia 42 (1986) 715–718.

18b Csaba, G., Receptor ontogeny and hormonal imprinting. Experientia 42 (1986) 750–759.

19 De Loof, A., New concepts in endocrine control of vitellogenesis and in functioning of the ovary in insects, in: Exogenous and Endogenous Influences on Metabolic and Neural Control, pp. 165–177. Eds A. D. F. Adding and N. Spronk. Pergamon Press, Oxford 1982.

20 Döhler, K. D., Development of hormone receptors conclusion. Experientia 42 (1986) 788–794.

21 Epple, A., Functional principles of vertebrate endocrine systems. Verh. dt. zool. Ges. (1982) 117–126.

22 Franke, H. D., and Pfannenstiel, H. D., Some aspects of endocrine control of polychaete reproduction. Fortschr. Zool. 29 (1984) 53–72.

23 Fitch, M. W., and Margoliash, E., Construction of Phylogenetic Trees. Science 155 (1967) 279–284.

24 Fraenkel, G., Interactions between Ecdysone, Bursicon and other endocrines during puparium formation and adult emergence in flies. Am. Zool. 15 (1975) 29–48.

25 Fraenkel, G., and Hsiao, C., Calcification, tanning, and the role of Ecdysone in the formation of the puparium of the facefly, Musca autumnalis. J. Insect Physiol. 13 (1967) 1387–1394.

26 Fujita, T., Yui, R., Iwanaga, T., Nishiistsutsuji-Uwo, J., Endo, Y., and Yanaihara, N., Evolutionary aspects of 'brain-gut' peptides: an immunochemical study. Peptides 2, suppl.2 (1981) 123–131.

27 Gorbman, A., Problems in the comparative morphology and physiology of the vertebrate. Thyroid gland, in: Comparative Endocrinology, pp. 266–282. Ed. A. Gorbman. Wiley, New York 1953.

28 Gorski, J., The nature and development of steroid hormone receptors. Experientia 42/7 (1986) 744–750.

29 Harvey, P. H., and Partridge, L., When deviants are favoured: Evolution of sex determination. Nature 307 (1984) 689–690.

30 Highnam, K. C., and Hill, L., The comparative endocrinology of the invertebrates, 2nd edn. Edward Arnolds, London 1977.

31 Hollenberg, M. D., Mechanisms of receptor-mediated transmembrane signalling. Experientia 42/7 (1986) 718–727.

32 Joosse, J., and Geraerts, W. P. M., Endocrinology, in: The Mollusca, vol. 4, pp. 317–406. Academic Press, London/New York 1983.

33 Karlson, P., Introduction: The concept of hormonal systems in retrospect and prospect. Nova Acta Leopoldina 56 (1984) 9–19.

34 Kende, H., and Gardner, G., Hormone binding in plants. A. Rev. Physiol. 27 (1976) 267–290.

166

35 Kleinholz, L. H., and Keller, R., Endocrine regulation in Crustacea, in: Hormones and Evolution I, pp. 159–213. Ed. E. J. W. Barrington. Academic Press, New York/San Francisco/London 1979.

36 Kochert, G., Sexual pheromones in algae and fungi. A. Rev. Plant Physiol. *29* (1978) 461–486.

37 Koller, G., Über Chromatophorensystem, Farbensinn und Farbwechsel bei Crangon vulgaris. Z. vergl. Physiol. *5* (1927) 191–246.

38 Koller, G., Versuche über die inkretorischen Vorgänge beim Garnelenfarbwechsel. Z. vergl. Physiol. *8* (1928) 601–621.

39 Kopec, S., Experiments on metamorphosis of insects. Bull. int. Acad. Cracovie (B) (1917) 57–60.

40 Kopec, S., Studies on the necessity of the brain for the inception of insect metamorphosis. Biol. Bull. *42* (1922) 323–342.

41 Kovacs, P., The mechanism of receptor development as implied by hormonal imprinting studies on unicellular organisms. Experientia *42* (1986) 770–775.

42 Marks, F., Molekulare Biologie der Hormone. Gustav Fischer Verlag, Stuttgart 1979.

43a Le Roith, D., Roberts, C. Jr, Lesniak, M. A., and Roth, J., Receptors for intercellular messenger molecules in microbes: Similarities to vertebrate receptors and possible implications for diseases in man. Experientia *42* (1986) 782–788.

43b Le Roith, D., Pickens, W., Wilson, G. L., Miller, B., Berelowitz, A. L., Collier, E., and Cleland, C. F., Somatostatin-like material is present in flowering plants. Endocrinology *117* (1985) 2093–2097.

44 Oksche, A., Evolution of neurosecretory cells and systems. Nova Acta Leopoldina NF *56*, 255 (1984) 39–50.

45 Pearse, A., The APUD concept and its verification by the use of molecular markers. Nova Acta Leopoldina NF *56*, 255 (1984) 177–124.

46 Pflugfelder, O., Entwicklungsphysiologie der Insekten, 2nd edn. Akad. Verl. Ges., Leipzig 1958.

47 Raabe, M., Insect Neurohormones. Plenum Press, New York/London 1982.

48 Romer, R., Ecdysteroids in Snails. Naturwissenschaften *66* (1979) 471–472.

49 Schraudolf, H., Action and phylogeny of antheridiogens. Proc. R. Soc., Edinburgh *86B* (1985) 75–80.

50 Schraudolf, H., personal communication.

51 Sehnal, F., Identical and similar hormones in various animal phyla: a case of chemical plesiomorphy. Evolution and Morphogenesis. Praha: Academia (1985) 507–517.

52 Slama, K., Gormonalnaja reguljacija morfogeneza u besponzvonocnych, evoljucionnye aspekty (Hormonal regulation of morphogenesis in invertebrates, evolutionary aspects). Z. obsc. Biol. *43* (1982) 805–822.

53 Wells, M. J., Hormonal control of reproduction in cephalopods, in: Perspectives in Experimental Zoology IV, pp. 157–166. Ed. P. Spencer Davies, 1976.

54 Werner H., Fridkin, M., Aviv, O., and Koch, Y., Immunoreactive and bioactive somatostatin-like material is present in tobacco (Nicotiana tabacum). Peptides *6* (1985) 797–802.

55 Wigglesworth, V. B., The Physiology of Insect Metamorphosis. Cambridge University Press, Cambridge 1954.

Receptors for intercellular messenger molecules in microbes: Similarities to vertebrate receptors and possible implications for diseases in man

D. LeRoith, C. Roberts Jr, M. A. Lesniak and J. Roth

Diabetes Branch, National Institute of Diabetes, and Digestive and Kidney Diseases, Bldg 10, Rm 8S-243, 9000 Rockville Pike, Bethesda (Maryland 20892, USA)

Outline and aim

Our focus is on the evolutionary origins of receptors for vertebrate hormones, neuroactive peptides, and related messengers. The first part will survey the possible evolutionary origins and phylogenetic distribution of the vertebrate-type messenger peptides providing a possible clue or guide to the same speculation for the receptor. Also, we will explain current data which suggest why receptors might need to be at least as old or as widely distributed as the messengers. In the latter part we will survey examples of materials in microbes that resemble vertebrate-type receptors and also highlight some possible applications to an understanding of human disease problems.

Introduction

A widespread pattern of intercellular communication involves release from one (secretory) cell a soluble messenger molecule which travels through an extracellular fluid compartment to another (target) cell (fig. 1).
The target cell has specific receptor molecules that recognize the messenger by binding it and the combination of messenger with receptor activates the target cell to yield a characteristic biological response.
This pattern of communication is well known for hormones, neuroactive substances, paracrine agents and pheromones in multicellular organisms e.g. verte-

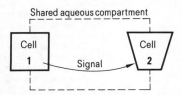

Figure 1. Basic features of intercellular communication systems. Cell 1 represents a secretory cell which is capable of synthesizing and releasing a signal messenger molecule into the shared fluid compartment. Cell 2, the target cell, contains a receptor which reacts to the signal messenger. In the endocrine system, cell 1 is a glandular cell and the messenger is a hormone; whereas in the nervous system, cell 1 is a neuron and the messenger is a neurotransmitter released into the synaptic space.

brates, invertebrates, and higher plants. Unicellular organisms including both eukaryotes (e.g. yeast) and prokaryotes (e.g. bacteria) utilize similar communication systems with soluble messengers designated pheromones to solve problems related to reproduction or nutrition[32, 40].

While communication systems with this overall design are very widespread, we may now ask how widespread are the molecules that mediate the communication? The peptide messenger molecules that are broadly represented among the vertebrates, e.g. hormones and neuroactive peptides, appear in many cases to be quite similar to substances found in a wide range of non-vertebrate animals[41, 18]. In some cases, the non-vertebrate forms appear to have essential biological functions in that organism.

In insects, insulin-related substances are present especially in neural elements[7, 21]. Injections of insulin accelerate glucose disposal, while surgical removal of the insulin-rich cells of the brain of the blowfly produces a hyperglycemic state, which can be ameliorated by insulin or extracts from insects that are rich in their own insulin-related materials[8]. Receptors for insulin are present in *Drosophila* which are similar to their vertebrate counterparts in binding properties, overall structure, and insulin-stimulated enzyme activity[34]. In molluscs, investigators have suggested a gastrointestinal tract site for the insulin-rich cells and have used anti-insulin antibodies to produce a catabolic ('diabetes-like') state[35].

We and others have found in microbes as well as in flowering plants, substances that closely resemble vertebrate messengers which suggest that these materials have early evolutionary origins and are distributed widely among different forms of life (tables 1 and 2)[1–3, 6, 10, 15, 20, 22–25, 29–31, 33, 42–44, 53]. Outside of the multicellular animals, no function of these vertebrate-type materials is known, although extreme conservation of the surface structure of these molecules including the receptor-reactive regions, suggests possibly (a) function and (b) binding to receptors. Thus the presence of the ligand raises the possibility of the presence of the homologous receptor, or at least that portion of the receptor that contains the specific ligand binding site.

Support for that suggestion is heightened by our current ideas of how hormones and related messengers and their respective receptors function in vertebrate

Table 1. Materials in unicellular organisms that resemble messenger peptides of vertebrates[1–3, 6, 20, 22, 24, 25, 29, 30, 33, 42, 43]

Hormone-related materials	Microbe
TSH	*Clostridium perfringens*
hCG	Many bacteria
Neurotensin	*E. coli, Caulobacter crescentus, Rhodopseudomonas palustrus*
Insulin	*Tetrahymena, E. coli, Neurospora crassa, Aspergillus fumigatus, Halobacterium solinarium, Bordetella pertussis*
Somatostatin	*Tetrahymena, E. coli, B. subtilis*
ACTH, β-endorphin	*Tetrahymena*
Relaxin	*Tetrahymena*
Arginine vasotocin	*Tetrahymena**
Calcitonin	*Tetrahymena, Candida, E. coli*

* Diabetes Branch unpublished observations.

Table 2. Materials in higher plants that resemble messenger peptides of vertebrates [10, 15, 23, 44, 52, 53]

Hormone-related material	Plant
LHRH	Oak leaves
TRH	Alfalfa
Opioid	Wheat
Interferon	Tobacco
Insulin	Spinach, Lemna, Rye*
Somatostatin -14 and -28	Spinach, Lemna, Tobacco

* Diabetes Branch unpublished data.

Table 3. Intrinsic biological properties of receptors for soluble messenger molecules

Function	Ligand
DNA-binding proteins	Steroid hormones; vitamin D-related sterols; thyroid hormones
Activator of G-proteins	Hormones and other messengers that activate (or inhibit) directly adenylate cyclase
Tyrosine-specific protein kinases	Epidermal growth factor (EGF); insulin; insulin-like growth factor-I (IGF-I); platelet derived growth factor (PDGF)
Ion channels	GABA-diazepam; acetylcholine (nicotinic)

systems. First, these messengers have no known function except as messengers. Secondly at the level of mechanism, the messenger peptides act solely to activate the receptor; the receptor is the proximate mediator of action at the target cell. Indeed in many systems we now know the particular kind of activity in the receptor that is turned on by the messenger (table 3).

Unicellular eukaryotes

Receptors for the mating factors in Saccharomyces cerevisiae. Saccharomyces cerevisiae is a budding yeast which exists in diploid form as well as two haploid forms, designated α and a (fig. 2). Union of the two haploids of opposite type is a process which requires two peptide pheromones or mating factors, α factor and 'a' factor, each produced by its namesake cell type to act on cells of the opposite type, via specific cell surface receptors[46]. (Interestingly, α factor bears a striking structural resemblance to the mammalian hypothalamic peptide known as GnRH or gonadotropin releasing hormone, can bind to the specific hormone binding site on the GnRH receptors of rat pituitary cells, and at high concentrations causes release of gonadotropins from the pituicytes)[14, 28].

More direct evidence for the existence of a receptor for the α mating factor has been reported by Jenness et al.[16]. These authors have demonstrated specific binding of α mating factor to 'a' cells and estimated that each cell has about 8000 binding sites. More recently, the nucleotide sequence of the gene coding for the α mating factor receptor has been determined and from it the amino acid sequence of the gene product (A. C. Burkholder, personal communication). A hydropathy analysis of the primary structure of the peptide reveals several

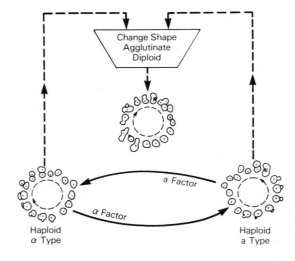

Figure 2. Sex pheromones in *Saccharomyces cerevisiae*. The life cycle of *S. cerevisiae* involves two mating type haploids (α and a) as well as sporulating diploids. Each haploid produces a pheromone (mating factor) that induces specific changes in the other haploid cell. This results in sexual conjugation with formation of diploid zygotes. Amino acid sequences of the N-terminus of α mating factor demonstrate significant homology with gonadotropin-releasing hormone of mammals (GnRH).

adjacent, evenly spaced, hydrophobic regions in the N-terminal 70% of the molecules, while the remainder of the molecule, i.e. the C-terminal portion, is largely hydrophilic. The hydrophobic regions probably represent the trans-membrane domains. Multiple transmembrane domains in receptors have also been demonstrated with rhodopsin[11] whereas typical vertebrate hormone receptors such as that for epidermal growth factor (EGF) and insulin have only one or two transmembrane domains per receptor complex[9, 49, 50]. In parallel studies, Sprague and colleagues have deduced the structure of the gene which is thought to encode the 'a' factor receptor and determined that the hydrophobic structure of this receptor is indeed similar to that described for the α factor receptor (G. Sprague, personal communication). Thus the mating system of *Saccharomyces cerevisiae* has several features in common with classic vertebrate hormone receptor systems, in that communication between the two haploid cells is brought about by soluble peptide messenger molecules and the receptors for these mating factors resemble vertebrate transmembrane proteins; in addition the α mating factor is structurally related to mammalian GnRH.

Another similarity between hormone-related systems in yeast and higher organisms involves the yeast adenylate cyclase system. The RAS1 and RAS2 genes of yeast are homologous to vertebrate ras oncogenes[47] and encode similar proteins. Genetic studies have shown that these RAS gene products are functional components of the yeast adenylate cyclase complex[36]. It has also been demonstrated that the yeast RAS proteins are membrane-bound and exhibit GTP-binding and hydrolyzing activities[48] involved in adenylate cyclase stimulation. Thus, the yeast RAS gene products are unicellular examples of the transducing proteins associated with vertebrate hormone-sensitive adenylate cyclase systems.

Estrogen binding in Saccharomyces. In the cytosol of *Saccharomyces cerevisiae* is a protein which selectively binds with high affinity a vertebrate estrogenic hormone, 17β-estradiol[4]. Unlabeled 17β-estradiol competes for the binding with tritiated estradiol while 17α-estradiol has only about 5% of the activity (fig. 3). Diethylstilbestrol, demoxaphin, nephoxidine and zearalenones, other natural and synthetic substances with estrogenic agonist or antagonist activity in mammals which bind strongly to the mammalian estrogen receptor are inactive in competing for a binding of tritiated estradiol by the *Saccharomyces* cytosolic protein. Despite this difference in the specificity of ligand binding, the yeast cytosolic protein shares physico-chemical properties with the estrogen receptor and other steroid hormone receptors of mammals. Feldman and co-workers also demonstrated that *Saccharomyces* contain lipid soluble materials that are capable of binding to the steroid binding site of estrogen receptors of mammals[4]. In addition, the yeast material can substitute for estrogen in stimulating uteri of ovariectomized mice. Thus, both ligand and receptor of yeast resemble similar components in mammals.

Corticosterone binding in Candida. Another yeast, *Candida albicans,* contains an intracellular protein that binds corticosterone (table 4). The specificity and affinity of binding of corticosterone to this yeast protein lies intermediate between those of two mammalian proteins, CBG, the cortisol binding globulin of plasma, and the glucocorticoid receptor, a cytosolic protein[26]. These investigators further suggest that the yeast probably produces an alternative ligand,

Table 4. Comparison of glucocorticoid binding proteins of *Candida albicans* and mammals

Property	*C. albicans* binding protein	Corticosterone binding globulin	Glucocorticoid receptor
Ligand	[^{3}H]corticosterone	[^{3}H]corticosterone	[^{3}H]dexamethasone
Binds synthetic glucocorticoids	No	No	Yes
Molecular weight	43,000	53,000	102,000
Hypertonic buffer	NEC4	3.4–4.1	4
Hypotonic buffer	NEC4	3.4–4.1	7–8
K_d(nM) (4°C)	7	1–7	3–31
K_{off} 4°C (min^{-1})	0.04	0.027	0.003
Stable at 37°C/1h	Yes	Yes	No

The properties of the *C. albicans* binding protein is closer to corticosterone binding globulin (CBG) of mammals than to the glucocorticoid receptor, but clearly not identical to either.

172

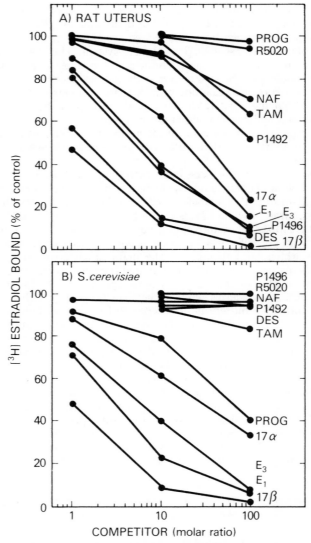

Figure 3. Competitive binding analysis of yeast estrogen binding protein and rat uterine estrogen receptor. A) Binding profile of [³H] estradiol in rat uterine cytosol. B) Binding profile in yeast. Competitors used; 17β-estradiol (17β); 17α-estradiol (17α); estrone (E₁); estriol (E₃); progesterone (Prog); tamoxifen (TAM); nafoxidine (NAF); diethylstilbestrol (DES); zearalanol (P1496); zearalenone (P1492); promegestone (R5020). (Reproduced from ref. 4).

possibly a lipid soluble substance which accumulates in the culture medium and has the ability to compete with labeled corticosterone for binding sites on the yeast protein.

Paracoccidioides brasiliensis. Paracoccidioides brasiliensis, a yeast that is pathogenic for humans, contains specific binding sites for labeled estradiol. Labeled steroid hormones including testosterone and corticosterone do not bind specifically to these yeast cells. The binding of labeled estradiol is of high affinity and can be competed for by estriol and progesterone with 25% of the affinity of estradiol; androgens, glucocorticoids, and the synthetic estrogen analog, diethylstilbestrol, have very low affinity[27]. Preliminary studies suggested that the physico-chemical characteristics of the binding protein are similar to those described for the steroid receptors of vertebrates and the binding site can be destroyed by trypsin. The transformation of the organism from mycelial to yeast form, which is an early step in the infection of humans can be inhibited by estradiol at concentrations similar to those encountered in vivo in a dose dependent fashion. These investigators further speculated that with paracoccidioidomycosis (formerly known as South American blastomycosis), the disease with is caused by this organism is much more pathogenic in men than in women despite apparently equal exposure to the organism because in women, the endogenous estrogen inhibits a key step in the yeast that leads to infection.

Sex pheromones and receptors in a water mold. Achlya, a unicellular water mold, has two distinct sex types, designated male and female (fig. 4). The female secretes a pheromone, antheridiol, which affects the male in many ways including an enhancement of secretion of another pheromone, oogoniol, which has as its target the female cell. Both antheridiol and oogoniol are exceedingly similar in structure to the classic steroid hormones of vertebrates. Investigators have found in *Achlya* a soluble intracellular protein that binds antheridiol specifically but not oogoniol or other more distantly related steroids including the steroid hormones of vertebrates. Only the male cells contain the antheridiol binding protein; the female cells lack it. This binding protein has several of the peculiar physical and chemical characteristics that are typical of steroid receptors of vertebrates (fig. 5a, b)[38, 39].

Opiate receptors in Amoeba. Feeding behavior or endocytosis of food particles by *Amoeba proteus* is highly regulated. Opioid alkaloids as well as opioid peptides in the nM range inhibit the endocytosis. This biological effect is antagonized by the bioactive stereospecific isomer of naloxone in a dose dependent manner, but not by its mirror image biologically inert form. These results suggest that the *Amoeba* has a specific receptor that elicits a biological response and has properties similar to those of the μ-type opioid receptor found in higher organisms. The existence of the receptor, its specificity, and its cellular connections have been demonstrated by these pharmacological experiments[17]. However, it has not yet been shown that this receptor acts physiologically to regulate normal feeding or other life processes in the organism. A similar approach has been used to detect other vertebrate-type receptors in the amoebae.

174

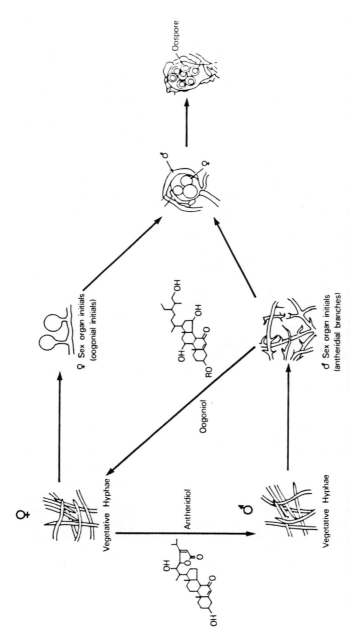

Figure 4. Sexual reproduction in the unicellular fungus, *Achlya ambisexualis*. Female continuously secretes antheridiol, a sterol pheromone, which induces in the male the formation of antheridial branches (sex organs) and secretion of oogoniol, a second pheromone, which causes the female to form sex organs. Antheridiol also attracts the male so it grows towards the female. Nuclei within both male and female sex organs undergo meiosis to form sperm and oospheres. A fertilization tube forms between male and female sex organs and male gametic nuclei pass into the oogonium. Sperm and oosphere fuse to produce zygotes ('oospores'). Under proper conditions, the zygotes then germinate and produce new diploid individuals. (Adapted from ref. 13).

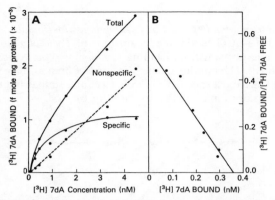

Figure 5a. Binding of [³H] 7dA to *Achlya*. Mycelia of *Achlya ambisexualis* were minced with scissors and homogenized using a polytron. Homogenates were filtered through glass wool and centrifuged at 250,000 × g for 2 h. The supernatant was clarified by filtration through 0.22 μm membrane. 286 μg aliquots of cytosol protein were incubated at 0 °C for 1 h in the presence of [³H]7dA, (a radiolabeled analogue of antheridiol) in the range of 0.1–4.5 nM. Similar incubations were performed in the presence of 50-fold molar excess of antheridiol. After the incubation period bound and free steroid were separated by dextran-coated charcoal. Non-specific binding is the amount of [³H]7dA bound in the presence of 50-fold excess of unlabeled antheridiol. Specific binding is calculated by subtracting non-specific from total at each concentration (A). Scatchard analysis (B) revealed an equilibrium dissociation constant of 0.65 nM and maximum binding capacity of 1245 fmoles/mg protein (Adapted from refs. 38 and 39 with permission).

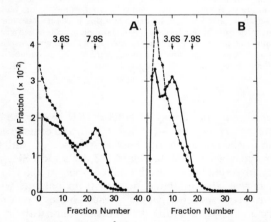

Figure 5b. Sucrose gradient analysis of [³H]7dA binding in *Achlya* cytosol. *Achlya* cytosol containing [³H]7dA, either with or without 200 nM antheridiol, were layered on 5–20 % w/w sucrose gradients and centrifuged for 16 h at 150,000 × g. Sedimentation profile in low ionic strength gradients containing molybdate (A) and high ionic strength gradients without molybdate (B). ●——● represents profile in presence of [³H]7dA alone and ●--- ● represents profile in the presence of a 50-fold excess of unlabeled antheridiol. These results were interpreted as suggesting that the *Achlya* cytosolic binding protein can be observed as apparent aggregated (8S) and dissociated (4S) states and highly sensitive to the stabilizing action of low ionic strength and sodium molybdate, as is typical of steroid hormone receptors of mammals. 3.6S arrow represents peak radioactivity from [¹⁴C] ovalbumin and 7.9S from [¹⁴C] aldolase. (Reproduced from ref. 38 with permission).

176

Prokaryotes

Chorionic gonadotropin binding sites in Pseudomonas maltophilia. Richert and Ryan found that ^{125}I-hCG binds to preparations of *Pseudomonas maltophilia* but not to other microorganisms, including *Pseudomonas aeruginosa* and other gram-negative rods[37]. The binding of the hCG was of high affinity with a single order of binding sites and a specificity which was quite similar but not identical with the hCG receptor of mammalian ovaries (fig. 6).

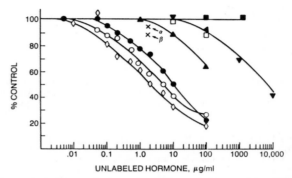

Figure 6. *Pseudomonas maltophilia* were grown to stationary phase in trypticase soy broth, sedimented by centrifugation, and resuspended in 40 mM Tris HCl, 10% sucrose. Bacterial pellets equivalent to 210 μg of bacterial protein were incubated with 2 ng of ^{125}I-hCG at 20°C for 20 h in the absence or presence of unlabeled hormones. Incubations were ended by precipitating bound radioactivity with carbowax and centrifugation at 2000 × g. (◊) human luteinizing hormone; (○) human chorionic gonadotropin; (●) ovine luteinizing hormone, (×) human chorionic gonadotropin α and β subunits; (▲) ovine follicular stimulating hormone; (□) ovine prolactin; (▼) ovalbumin, ovomucoid, bovine gamma globulin; (■) sugars (glucose, galactose, mannose, maltose, D-glucosamine, N-acetyl-neuraminic acid); (☐) KI. (Reproduced from ref. 37 with permission).

TSH binding proteins in bacteria. Using ^{125}I-labeled TSH, Weiss and co-workers detected binding sites in *Yersinia enterocolitica, E. coli,* and other gram-negative rods[51]. The binding of the labeled hormone was specific, i.e., unlabeled TSH competed better for the labeled TSH than did hCG, LH, and FSH, all of which are related to TSH; unrelated hormones did not compete at all. The binding of TSH to the bacteria was also inhibited by sera from patients with Graves' disease but not from sera from other people[12]. Since Graves' disease is thought to be etiologically caused by autoantibodies directed against the receptors for TSH on the thyroid gland, these investigators were able to show that enrichment of these antibodies inhibited TSH binding to the binding sites on the gram-negative organisms (fig. 7). These authors suggested the possibility that, in certain individuals, bacterial products (i.e., TSH-binding sites) may generate antibodies which cross-react with the endogenous TSH receptors of the thyroid gland. These antibodies in the absence of TSH may activate the thyroid cells causing them to overproduce their hormonal products resulting in hyperthyroidism in these patients.

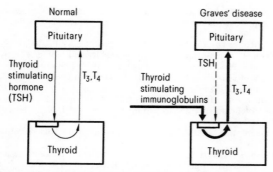

Figure 7. Hyperthyroidism associated with Graves' disease in humans is thought to be caused by 'autoantibodies' directed against the TSH receptor of the thyroid cells. *Yersinia enterocolitica* and *E. coli* have TSH-receptor-like material that binds ^{125}I-TSH in vitro, and this binding is inhibited by unlabeled TSH. Sera of patients with Graves' disease also inhibit the binding of labeled TSH to the bacteria. These studies have suggested that the TSH binding protein of bacteria and that of human thyroid membranes have homologies. Additional evidence is the production of anti-TSH receptor antibodies by rabbits injected with *Yersinia enterocolitica*.

Interestingly, *E. coli, Proteus vulgaris* and *Klebsiella pneumoniae* share antigenic determinants with the nicotinic acetylcholine receptor and those authors have postulated a possible role of these bacteria in the pathogenesis of *Myasthenia gravis*, a disease causally related to autoantibodies directed against the acetylcholine receptor[45]. Similarly, acute rheumatic fever follows infection with *Streptococcus pyogenes*, and interestingly murine monoclonal antibodies to *Streptococcus pyogenes* react with skeletal muscle myosin[19].

Conclusion

Intercellular communication analogous to that found among vertebrates plays an important biological role in representative organisms at all levels of life, including microbes. Among the small number of well-characterized intercellular messengers indigenous to microbial systems, several are structurally similar to messenger molecules of vertebrates. In addition, the molecules of intercellular communication typical of vertebrate systems – the intercellular messengers, their receptors, as well as their post-receptor intracellular components – show similarities to molecules of both eukaryotic and prokaryotic microbes; the normal function of these microbial components is, in most cases, not yet known. Overall, it appears that the extent of the overlap in the field of intercellular communication between vertebrates and non-vertebrates, as well as between multicellular organisms and unicellular organisms, is much more extensive than heretofore suspected.

178

1 Acevedo, M. F., Slifkin, M., Pouchet, G. R., and Pardo, M., Immunocytochemical localization of a choriogonadotropin like protein in bacteria isolated from cancer patients. Cancer *41* (1978) 1217–1219.
2 Berelowitz, M., LeRoith, D., Von Schenk, H., Newgard, C., Szabo, M., Frohman, L. A., Shiloach, J., and Roth, J., Somatostatin-like immunoactivity and biological activity is present in T. pyriformis, a ciliated Protozoan. Endocrinology *110* (1982) 1939–1944.
3 Bhatnagar, Y. M., and Carraway, R., Bacterial peptides with C-terminal similarities to bovine neurotensin. Peptides *2* (1981) 51–59.
4 Burshell, A., Stathis, P. A., Do, Y., Miller, S. C., and Feldman, D., Characterization of an estrogen-binding protein in the yeast Saccharomyces cerevisiae. J. biol. Chem. *259* (1984) 3450–3456.
5 Csaba, G., The present state in the phylogeny and ontogeny of hormone receptors. Horm. Metab. Res. *16* (1984). 329–335.
6 Deftos, L., LeRoith, D., Shiloach, J., and Roth, J., Salmon calcitoninlike immunoactivity in extracts of Tetrahymena pyriformis. Horm. Metab. Res.*17* (1985) 82–85.
7 Duve, H., and Thorpe, A., Immunofluorescent localization of insulin-like material in the median neurosecretory cells of the blowfly Calliphora vomitoria (Diptera). Cell Tissue Res. *200* (1979) 187–191.
8 Duve, H., Thorpe, A., and Lazarus, N. R., Isolation of material displaying insulin-like immunological and biological activity from the brain of the blowfly, Calliphora vomitoria. Biochem. J. *184* (1979) 221–227.
9 Ebina, Y., Ellis, L., Jarnagin, K., Edery, M., Graf, L., Clauser, E., Ou, J.-H., Maslarz, F., Kan, Y. W., Goldfine, I. D., Roth, R. A., and Rutter, W. J., The human insulin receptor cDNA: The structural basis for hormone-activated transmembrane signalling. Cell *40* (1985) 747–758.
10 Fukushima, J., Wanatabe, S., and Kaushima, K., Extraction and purification of substance with luteinizing hormone releasing activity from the leaves of avenasativa. Tohuku J. expl Med. *119* (1976) 115–122.
11 Hargrave, P., Progress in Retinal Research, vol. 1, pp. 1–40, 1982.
12 Heyma, P., Harrison, L., and Robins-Browne, R., submitted (1986).
13 Horgen, P. A., The role of the steroid sex pheromone antheridiol in controlling the development of male sex organs in the water mold, Achlya, in: Sexual Interactions in Eukaryotic Microbes, pp. 155–178. Eds D. H. O'Day and P. A. Horgen. Academic Press, New York 1981.
14 Hunt, L. T., and Dayoff, M. D., Structural and functional similarities among hormones and active peptides from distantly related eukaryotes, in: Peptides: Structure and Biological Function. Proceedings of the Sixth American Peptide Symposium, pp. 757–760. Eds E. Gross and J. Meienhofer. Pierce Chemical Co. Rockford, Illinois 1979.
15 Jackson, I. M. D., Abundance of immunoreactive thyrotropin-releasing hormone-like material in the alfalfa plant. Endocrinology *108* (1981) 344–346.
16 Jenness, D. D., Burkholder, A. C., and Hartwell, L. H., Binding of α-factor pheromone to yeast cells: Chemical and genetic evidence for an α-factor receptor. Cell *35* (1983) 521–529.
17 Josefsson, J. O., and Johansson, P., Naloxone-reversible effects of opioids on pinocytosis in amoebe proteus. Nature *282* (1979) 78–80.
18 Kramer, K.J., Vertebrate hormones in insects, in: Comprehensive Insect Physiology, Biochemistry and Pharmacology, vol. 7. Endocrinology, in press (1985).
19 Krisher, K., and Cunningham, M. W., Myosin: A link between Streptococci and heart. Science *227* (1985) 413–415.
20 LeRoith, D., Berelowitz, M., Pickens, W., Crosby, L. K., and Shiloach, J., Somatostatin-releated material in E. coli: Evidence for two molecular forms. Biochim. biophys. Acta *838* (1985) 355–340.
21 LeRoith, D., Lesniak, M. A., and Roth, J., Insulin in insects and annelids. Diabetes *30* (1981) 70–76.
22 LeRoith, D., Liotta, A. S., Roth, J., Shiloach, J., Lewis, M. E., Pert, C. B., and Krieger, D. T., Corticotropin and β endorphin-like materials are native to unicellular organisms. Proc. natn. Acad. Sci. USA *79* (1982) 2086–2090.
23 LeRoith, D., Pickens, W., Wilson, G. L., Miller, B., Berelowitz, M., Vinik, A. I., Collier, E., and Cleland, C. F., Somatostatin-like material is native to flowering plants. Endocrinology *117* (1985) 2093–2097.
24 LeRoith, D., Shiloach, J., Roth, J., and Lesniak, M. A., Evolutionary origins of vertebrate hormones: Substances similar to mammalian insulins are native to unicellular organisms. Proc. natn. Acad. Sci. USA *77* (1980) 6184–6188.

25 LeRoith, D., Shiloach, J., Roth, J., and Lesniak, M. A., Insulin or a closely related molecule is native to Escherichia coli. J. biol. Chem. *256* (1981) 6533–6536.

26 Loose, D. S., and Feldman, D., Characterization of a unique corticosterone-binding protein in Candida albicans. J. biol. Chem. *257* (1982) 4925–4930.

27 Loose, D. S., Stover, E. P., Restrepo, A., Stevens, D. A., and Feldman, D., Estradiol binds to a receptor-like cytosol binding protein and initiates a biological response in Paracoccidioides brasiliensis. Proc. natn. Acad. Sci. USA *80:* (1983) 7659–7663.

28 Loumaye, E., Thorner, J., and Catt, K. J., Yeast mating pheromone activates mammalian gonadotrophs: Evolutionary conservation of a reproductive hormone. Science *218* (1982) 1324–1325.

29 Macchia, V., Bates, R. W., and Pastan, I., Purification and properties of thyroid stimulating factor isolated from Clostridium perfringens. J. biol. Chem. *242* (1967) 3726–3730.

30 Maruo, T., Cohen, H., Segal, S. J., and Koide, S. S., Production of choriogonadotropin-like factor by a microorganism. Proc. natn. Acad. Sci. USA *76* (1979) 6622–6626.

31 Morley, J. E., Meyer, N., Pekary, A. E., Melmed, S., Carso, H. E., Briggs, J. E., and Hershman, J. M., A prolactin inhibitory factor with immunocharacteristics similar to thyrotropin releasing factor (TRH) is present in rat pituitary tumors (GH3&W5) testicular tissue and a plant material, Alfalfa. Bioch. biophys Res. Comm. *96* (1980) 47–53.

32 O'Day, D. H., Modes of cellular communication and sexual interactions in eukaryotic microbes, in: Sexual Interactions in Eukaryotic Microbes, pp. 3–17. Eds D. H. O'Day and P. A. Horgen. Academic Press, New York 1981.

33 Perez-Cano, R., Murphy, P. K., Girgis, S. I., Arnett, T. R., Blankharn, I., and MacIntyre, I., Unicellular organisms contain a molecule resembling human calcitonin. Endocrinology *110* (1982) 673 (Abstract).

34 Petruzzelli, L., Herrera, R., Garcia, R., and Rosen, O. M., The insulin receptor of Drosophila melanogaster. Cold Spr. Harb. Symp. *3* (1985) 115–121.

35 Plisetskaya, E., Kazakov, V. K., Solititskaya, L., and Leibson, L. G., Insulin producing cells in the gut of freshwater bivalve molluscs Anodonta cygnea and Unio pictorum and the role of insulin in the regulation of their carbohydrate metabolism. Gen. comp. Endocr. *35* (1978) 133–145.

36 Powers, S., Katauka, T., Fasano, I., Goldfarb, M., Strathern, J., Broach, J., and Wigler, M., Genes in S. cerevisiae encoding proteins with domains homologous to the mammalian ras proteins. Cell *36* (1984) 607–612.

37 Richert, N. D., and Ryan, R. J., Specific gonadotropin binding to Pseudomonas maltophilia Proc. natn. Acad. Sci. USA *73* (1977) 878–882.

38 Riehl, R. M., and Toft, D. O., Analysis of the steroid receptor of Achlya ambisexualis. J. biol. Chem. *259* (1984) 15324–15330.

39 Riehl, R. M., Toft, D. O., Meyer, M. D., Carlson, G. L., and McMorris, T. C., Detection of a Pheromone-binding protein in the aquatic fungus Achlya ambisexualis. Expl Cell Res. *153* (1984) 544–549.

40 Roth, J., LeRoith, D., Shiloach, J. Rosenzweig, H. L., Lesniak, M. A., and Havrankova, J., The evolutionary origins of hormones, neurotransmitters, and other extracellular chemical messengers. New Engl. J. Med. *306* (1982) 523–527.

41 Roth, J., LeRoith, D., Shiloach, J., and Rubinovitz, C., Intercellular communication: An attempt at a unifying hypothesis. Clin. Res. *31* (1983) 354–363.

42 Rubinovitz, C., and Shiloach, J., Insulin-related material in prokaryotes. FEMS Microbiol. Lett., in press (1986).

43 Schwabe, C., LeRoith, D., Thompson, R. P., Shiloach, J., and Roth, J., Relaxin extracted from protozoa (Tetrahymena pyriformis). J. biol. Chem. *258* (1983) 2778–2782.

44 Sela, I., Plant-virus interaction related to resistance and localisation of viral infections. Adv. Virus Res. *26* (1981) 201–237.

45 Stefansson, K., Dieperink, M. E., Richman, D. P., Gomez, C. M., and Marton, L. S., Sharing of antigenic determinants between the nicotinic acetylcholine receptor and proteins in Escherichia coli, Proteus vulgaris and Klebsiella pneumoniae: Possible role in the pathogenesis of Myasthenia gravis. New Engl. J. Med. *312* (1985) 221–225.

46 Stotzler, D., and Duntze, W., Isolation and characterization of four related peptides exhibiting α factor activity from Saccharomyces cerevisiae. Eur. J. Biochem. *65* (1976) 257–262.

47 Temeles, G., Gibbs, J., D'Alonzo, J. S., Sigal, J. S., and Scolnick, E., Yeast and mammalian ras proteins have conserved biochemical propterties. Nature *313* (1985) 700–703.

48 Toda, T., Uno, I., Ishikawa, T., Powers, S., Kataoka, T., Broek, D., Cameron, S., Broach, J., Matsumoto, K., and Wigler, M., In yeast, RAS proteins are controlling elements of the cyclic AMP effector pathway. Cell *40* (1985) 27–36.

49 Ullrich, A., Bell, J. R., Chen, E. Y., Herrera, R., Petruzzelli, L. M., Dull, T. J., Gray, A., Coussens, L., Liao, Y.-C., Tsubokawa, M., Mason, A., Seeberg, P. H., Grunfeld, C., Rosen, O. M., and Ramachandran, J., Human insulin receptor and its relationship to the tyrosine kinase family of oncogenes. Nature *313* (1985) 756–761.

50 Ullrich, A., Coussens, L., Hayflick, J. A., Dull, T. J., Gray, A., Tam, A. W., Lee, J., Yarden, Y., Liberman, T. A., Schlessiner, J., Downward, J., Mayes, E. L. V., Whittle, N., Waterfield, M. D., and Seeburg, P. H., Human epidermal growth factor receptor cDNA sequence and aberrant expression of the amplified gene in A431 epidermoid carcinoma cells. Nature *309* (1984) 418–425.

51 Weiss, M., Ingbar, S. H., Winblad, S., and Kasper, D. L., Demonstration of a saturable binding site for thyrotropin in Yersinia enterocolitica. Science *219* (1983) 1331–1333.

52 Werner, H., Fridkin, D. A., and Koch, Y., Immunoreactive and bioactive somatostatin-like material is present in tobacco. Peptides, in press (1986).

53 Zioudrov, C., Streaty, R. A., and Klee, W. A., Opioid peptides derived from food proteins: The exorphins. J. biol. Chem. *254* (1979) 2446–2450.

Development of hormone receptors: Conclusion

K. D. Döhler

Bissendorf Peptide GmbH, Burgwedeler Str. 25, D–3002 Wedemark 2 (Federal Republic of Germany)

When I was asked to write the conclusion to this multi-author review, I was somewhat reluctant at first, since the biochemistry of hormone receptors is far from my own field of research. However, I accepted this invitation because it is my personal experience that the view across borders into neighboring fields can bring new insight into one's own field of endeavor. I apologize if my sometimes unorthodox way of thinking about scientific problems and questions may now and then challenge traditional points of view, but possibly this is the true reason why I was asked to write this concluding review. Scientific data are traditionally looked at with the attitude that *anything that cannot be proved does not exist.* Historical development, however, has repeatedly shown that the attitude that *only those things which can be disproved do not exist* may hit the truth much better.

After reading these highly interesting expert reviews on specialized questions of hormone and receptor structure and function, phylogeny and ontogeny, imprinting and recycling, it would be most presumptuous of me to claim to present a final answer to the many open questions on how receptors may arise. Therefore, all those who expect the elaboration of a textbook-style theory on receptor development, based on classical points of view, will be disappointed. Research on receptor development is not ready yet for a classical theory. Classical theories give the impression that everything is already known and, owing to this impression, they inhibit the development of new insights and further research. *Man makes theories, but nature does as it very well pleases.* And because nature does as it pleases, a conclusion is nothing else but *the point where one gets tired of thinking.*

Even in science it is easier to swim with the stream (which scientist has not yet had this experience, especially when making a grant application?), but if you want to get to the source then you must frequently swim against the stream. So please forgive me, if I sometimes leave the treadmill of traditional thinking during my vague attempt to tie up some of the open ends. Remember, *you cannot pass someone, if you only keep stepping into his footsteps.*

Vertebrate hormones exist in invertebrates, in unicellular organisms and in plants

The cells of the body communicate with each other through biochemical mediators, like hormones, neuroactive substances, paracrine agents and pheromones. Previous reviews[3, 11, 35, 42] have demonstrated that this particular method of communication existed already very early in evolution and has survived for hundreds of millions of years. Throughout these years a great number of mutations have generated new phyla, new classes, new orders and an enormous number of species with immense interindividual variability.

Is it not surprising how stable the structures of biochemical mediators are? They have survived millions of mutations from unicellular organisms to mankind. Vertebrate hormones have been detected in non-vertebrate animals, in plants, and even in unicellular organisms. Insulin, for example, has been identified in insects[19, 38], annelids[38], and molluscs[57], in a variety of unicellulars[43, 44, 60], and in plants like spinach, lemna and rye (for a review see ref. 42). Neurotensin[5], somatostatin[4, 39], ACTH[40], β-endorphin[40], relaxin[62], and calcitonin[16, 55] have been identified in various unicellular organisms.

Although the universal distribution of these hormones has been well documented, very little is known yet about the function of vertebrate hormones in lower organisms. Insulin might universally be involved in carbohydrate metabolism; its involvement has been demonstrated not only for vertebrates, but also for insects[56] and molluscs[57]. Similarly, the pituitary hormone ACTH and the opioid β-endorphin may not only be involved in the stress response of vertebrates, but also in the stress response of other organisms, since the stress response is a very universal phenomenon for adaptation. Then how about the function of opioids in wheat[70]? Is it imaginable that plants respond to stress with a hormone reaction similar to that of animals? The identification of interferon in tobacco[63] raises the question whether plants may respond to viral infections in the same way as animals. With the best efforts of my imagination, however, I cannot suggest a function for the pregnancy hormone human chorionic gonadotropin (hCG), which has been identified in bacteria[1, 49], or for luteinizing hormone-releasing hormone (LHRH), which has been extracted from oak leaves[21], not to mention the presence of somatostatin in spinach, lemna and tobacco[41] and the presence of thyrotropin-releasing hormone (TRH) in alfalfa vegetable[27, 51].

Lower organisms possess receptors for vertebrate hormones

What is the functional significance of vertebrate hormones in lower organisms? This question will certainly give rise to some very interesting research projects in the future, but at present we lack possible answers. Instead, for the time being, we must be content to search for answers to the question whether or not lower organisms actually possess binding sites or receptors for vertebrate hormones. This question was reviewed by LeRoith et al.[42]. The authors listed a number of examples of the existence of receptors for vertebrate hormones in lower organisms. Estrogen receptors and corticosterone receptors were observed in different species of yeast[8, 45, 46], opiate receptors were localized in amoeba[30], and binding sites for TSH were observed in a number of bacterial species[68]. The presence of receptors for vertebrate hormones in non-vertebrate organisms supposes that these receptors may have physiological meaning for these organisms. In addition, these receptors may be of important clinical relevance. Infection through bacteria, which carry receptors for human hormones, may generate an immune response in humans against the bacterial receptors, and may then encroach upon the inherent endogenous human receptors. Heyma,

Harrison and Robins-Browne (cited by ref. 42) suggested the possibility that bacterial TSH-binding sites may generate antibodies in humans which cross-react with the endogenous TSH receptors of the human thyroid gland. These antibodies may activate the production of thyroid hormones in the thyroid cells, which will result in hyperthyroidism. The authors observed that binding of TSH to TSH-receptor carrying bacteria was inhibited by sera from patients with Graves disease, but not by sera from other people. Thus, the study of endocrine phylogeny is not only of comparative interest; it may be of great importance for the clinical endocrinologist as well.

The multifold existence of receptors for vertebrate hormones in lower organisms of different phyla gives rise to the question whether actually each individual organism in this universe may already possess the whole array of hormone receptors known to exist in vertebrates. The fact that binding sites for human chorionic gonadotropin were observed in *Pseudomonas maltophila,* but not in other microorganisms, such as *Psdeudomonas aeruginosa* and other gram negative rods[59], indicates that the development of a particular hormone receptor is a very specific affair for each individual species (or individual organism?). On the other hand, it was already suggested previously[18] that sometimes methodological problems may mask the true results. Thus, *the inability for a particular method to detect a certain parameter is no certain proof of the non-existence of this parameter.* Frowein et al.[20] demonstrated that gonadotropin receptors exist in gonadotropin-insensitive Leydig cells of immature rats; however, they are present in a 'masked' form, and they bind to gonadotropins only after they have been 'unmasked'.

Evolutionary aspects of hormones and receptors

The fact that vertebrate hormones and their receptors have been detected in lower organisms does not necessarily mean that their chemical and physical structure as well as their biological activity is fully identical in all organisms from unicellulars to humans. True comparisons are usually difficult to make, because of the profound differences in organization and mode of life between widely separated taxa[2]. It was mentioned previously that some of the same peptide hormones can be detected in vertebrates, invertebrates, unicellulars and even plants. The chemical and physical structure of peptide and polypeptide hormones, however, has in many cases undergone diversification throughout phylogeny. These diversifications can arise at the level of the genetic code by point mutations within the DNA, resulting in specific amino acid replacements within peptide hormones[3]. Insulin molecules, for example, differ in different species with respect to the number and position of amino acid substitutions. After 500 million years of independent evolution the insulin molecule of the hagfish differs from that of man in approximately 38% of its amino acids[52]. Nevertheless, the different insulin molecules of different species share common biological properties[66].

Diversification of peptide molecules throughout evolution raises the question whether or not the receptors for these hormones may, in parallel, have undergone equivalent diversification. This question has been studied in detail by extraction and characterization of hyperglycemic hormones from several crustaceans, including an isopod and two decapods[48]. These hormones were shown to be peptides with 50–58 amino acids, with overall similarity of composition, but with much interspecies variation in amino acid substitutions. Cross-reaction studies indicated that the receptors have varied side by side with the hormones[3, 48].

As mentioned in the previous paragraph, hormones may undergo structural alterations during evolution without changing their biological properties. As reviewed in detail by Barrington[3] hormones of common ancestry may also undergo functional diversification during evolution and may eventually establish new target relationships. Such a case was discussed by Barrington[3], concerning the structural resemblance of 4 invertebrate peptide hormones with different biological functions. The 8-residue erythrophore-concentrating hormone of crustaceans, the 10-residue adipokinetic hormone of insects and the two 8-residue peptides periplanetin CC-1 and periplanetin CC-2, which both show adipokinetic activity in grasshoppers and hyperglycemic activity in cockroaches[61], are sufficiently alike structurally to imply divergence by point mutation from a single ancestral molecule[3, 50]. Similarly, the two neurohypophyseal peptide hormones vasopressin and oxytocin have different biological functions, although they are both derived from the common ancestor molecule arginine vasotocin. Arginine vasotocin exists in fish, amphibians, reptiles, and birds, but not in mammals. Finally, a particular hormone from one species may be totally inactive in another species. Human growth hormone, for example, stimulates growth in humans and monkeys. Bovine growth hormone stimulates growth in the bovine, but it is ineffective in humans and monkeys. Similarly, mammalian gonadotropic hormones, even in tremendous doses, do not stimulate the gonads of certain amphibian species, whereas relatively small amounts of amphibian pituitary extracts will do so[22].

Sometimes a pedigree of endocrine phylogeny can be established. As reviewed by Barrington[3], the two mammalian pituitary hormones somatotropin (growth hormone) and prolactin have some 23% of their residues in common, which is generally conceded to indicate divergence from a common molecular ancestor. Immunological evidence has shown that growth hormone and prolactin are already present in the lamprey's pituitary gland, at the beginning of vertebrate evolution; this indicates that the two hormones must have diverged very early during evolution. A third member of this hormone family, human placental lactogen (hPL), shares some 85% of its amino acid residues with human growth hormone. This great similarity indicates that hPL must have diverged from growth hormone much later, probably during primate evolution[67]. During the long period of prolactin evolution, this hormone has established a great variety of target relationships. Just to mention a few examples, prolactin maintains luteal function and initiates milk secretion in mammals, it stimulates broody behavior and the production of crop milk in birds, it stimulates behavioral water drive in terrestrial salamanders, and it stimulates nest-building, melanogenesis and osmoregulation in some fishes[22].

The genetic code for receptor synthesis

We have seen from previous reviews[3, 11, 35, 42] that there are hormones which existed very early in evolution and which remained structurally and functionally rather stable. Other hormones underwent structural modifications, but their biological functions remained the same. A third group of hormones remained structurally rather stable, but their biological property changed during evolution. The change in biological activity sometimes went in parallel with the necessity for the species to adapt to a new environment, or with the phylogenetic development of new organ systems, which then became the new target for an old hormone, phylogenically speaking. Some hormones, finally, have changed during evolution in both structure and function.

In any case, the evolutionary development of a new hormone, or the mutational alteration of an already existing hormone, must go in parallel with the development of a new specific receptor, or with the structural adaptation of an already existing receptor. This new receptor must be able to recognize and interact with the new hormone in order to elicit a biological function. Teleologically speaking, it would be most efficient for nature if the same mutation which generates a new hormone could elicit a complementary mutation, which provides the specific receptor.

In this regard an interesting pattern in the genetic code was recently observed. Codons for hydrophilic and hydrophobic amino acids on one strand of nucleic acid are complemented by codons for hydrophobic and hydrophilic amino acids on the other strand[6]. The average tendency is for codons for 'charged' amino acids to be complemented by codons for 'uncharged' amino acids. Following this pattern, two peptides that represent complementary strands of nucleic acid will display an interchange of their hydrophilic and hydrophobic amino acid residues when the amino terminus of one peptide is aligned with the carboxyl terminus of the other. For receptor-hormone interaction and binding to occur, the two structures have to be complementary with respect to their hydrophobic and hydrophilic domains. One possible consequence of this observation is suggested by the finding that many biologically important peptides, composed of 10 to 50 amino acids, can assume amphiphilic secondary structures in the presence of another amphiphilic structure such as a membrane or a receptor binding site[31]. As a result of this relationship Bost et al.[7] hypothesized that complementary DNAs, when transcribed in the 5′ to 3′ direction and in the same reading frame, will code for peptides or proteins that interact. This hypothesis was tested by the authors, and they demonstrated that the two naturally-occurring peptides corticotropin (ACTH) and gamma-endorphin bind specifically and with high affinity to synthetically derived counterpart peptides, which had been specified by RNA sequences complementary to the mRNA for ACTH and gamma-endorphin, respectively[7].

DNA molecules resemble chain ladders, twisted into a helix, in which pairs of bases join two linear chains constructed from deoxyribose and phosphate subunits. The bases invariably pair so that adenine links to thymine and guanine links to cytosine on the complementary DNA strands. Adenine, thymine (which is replaced on RNA by uracil), guanine and cytosine act as code letters, ultimately for the recognition of particular amino acids during ribosomal peptide synthesis. Due to the complementary structural relationship between the two DNA strands, a mutation on one strand would inevitably elicit a complementary mutation on the other strand. In other words, a point mutation occurring in a particular section of one DNA strand, which encodes for a particular peptide hormone, will result in the substitution of one amino acid in this particular peptide molecule. If the substituted amino acid has different hydrophobicity or hydrophilicity as compared to the previous amino acid, then the peptide will assume a secondary and tertiary structure different from that of the premutational peptide. Since the point mutation has altered one partner of one pair of bases, let us say adenine was replaced by cytosine, then thymine as the other partner of this base-pair must be replaced by guanine in order to ensure cross-linkage of the two complementary DNA strands. As a natural consequence the transcription of the mutated complementary DNA strand would encode for the synthesis of a peptide receptor which has acquired one substituted amino acid. This substitution may determine the secondary and tertiary structure of this new receptor molecule in such a way that it can interact with the new peptide hormone.

Another way in which evolution may have generated the primordial receptor of a peptide is by translocation of the peptide's complementary DNA sequence.

Hormone-induced activation of receptor synthesis

In every multicellular organism the genetic code of this organism is present in each one of its individual cells. Activation of the genetic code, however, must occur differently in different cells, because during cell differentiation each cell is known to acquire its very specialized structure and function. Transcription of both strands of DNA of a particular gene in the same reading frame is known not to take place within the same cell. Considering receptor-hormone interaction, it would not make sense, teleologically speaking, if the hormone and its receptor were to be synthesized within the same cell. A hormone is, by definition, a chemical mediator which delivers information to distant cells, via the extracellular fluid. If the hormone and its receptor were synthesized within the same cell, then the specific affinity of the receptor for the hormone would not allow the hormone to leave the cell. Therefore, a mechanism must be in operation, which activates transcription of a particular DNA sequence in one group of cells (i.e. the endocrine gland) and activates transcription of the complementary DNA sequence in another group of cells (i.e. the hormonal target organ). This mechanism would allow for specific cellular recognition and for distant communication via the interacting peptide products.

The mechanisms by which transcription and translation of the genetic code are activated, are still very obscure and leave plenty of room for speculation. Hormones and/or receptor-hormone complexes may very well be involved in activation of the genetic code. DNA binding by steroid receptors has been a popular topic for many years[69]. After activation by estrogens, estrogen receptors are known to bind tightly to DNA[24]. As mentioned in the review by Gorski[23] it was shown recently that progesterone receptors[53] and glucocorticoid receptors[32, 54] bind to very specific sequences of DNA, which are known to be essential for the expression of progesterone and glucocorticoid action, respectively.

Estrogenic interaction with the genetic material in the cell nucleus was shown to stimulate DNA synthesis and estrogen receptor resynthesis in the cytosol[28, 29]. Estrogens were also shown to stimulate the synthesis of progesterone receptors[17, 33, 34]. The time course of this induction suggests that a transcriptional site of action is involved[23]. Posner et al.[58] observed that estrone and estradiol induced lactogenic receptors in rat liver membranes via a pituitary mechanism. The receptors were able to bind ovine prolactin and human growth hormone with high affinity. Hypophysectomy drastically decreased the levels of lactogenic receptors in mature female rats and in these animals estrogen failed to restore receptor levels[58]. Thus, a factor released by the pituitary gland (growth hormone?, prolactin?) is primarily responsible for the induction of lactogenic receptors.

Hormone-induced organization of receptors

A great number of studies, performed with unicellular organisms, seems to indicate that the extracellular hormonal environment of a cell can determine the development of receptor systems in this particular cell[11, 35]. Serotonin and histamine, the phagocytosis-stimulating hormones of higher organisms, also stimulate phagocytosis of the ciliated unicellular organism *Tetrahymena*[13]. Insulin and adrenalin stimulate glucos metabolism, thyroxine stimulates growth, and thyrotropin (TSH), follicle stimulating hormone (FSH) and ACTH stimulate RNA synthesis in *Tetrahymena*[10]. The first encounter of *Tetrahymena* with the hormones mentioned above was shown to enable the organism to bind these hormones to a greater extent on the next occasion[36], and the response of this organism to the particular hormone increased on reexposure[9]. Antibodies raised against rat hepatocellular insulin receptors were shown to bind to *Tetrahymena* cells which had been pretreated with insulin, to a significantly greater extent than to untreated control cells[12]. From these studies, it appears that the first encounter of *Tetrahymena* cells with insulin generated insulin-receptors which were immunologically similar to the insulin receptors of the rat hepatocellular membrane. This conclusion is supported by the observation that treatment of rat hepatocytes with antiserum, raised against insulin-pretreated live *Tetrahymena* cells, inhibited the insulin binding capacity of rat hepatocytes to a greater extent than antiserum raised against untreated *Tetrahymena* cells[37]. The authors concluded that the insulin receptors of the *Tetrahymena* are not preformed structures, but seem to arise under the influence of insulin[35].

As the result of a great number of studies with various hormones, Csaba and his associates came to the conclusion that hormonal presence is an indispensable prerequisite of receptor formation[9, 11, 35]. The first encounter of a particular hormone with a primordial binding site will result in the 'imprinting' of this binding site into the complementary shape of the hormone. This mechanism of imprinting will permanently increase the binding affinity of the primordial binding site to the hormone. The imprinted receptor will 'memorize' the shape of the hormone throughout many cell divisions. In *Tetrahymena* this 'memory' for the binding of insulin was still demonstrable after as many as 500 cell generations[11]. However, this 'memory' will eventually fade if the receptor is deprived of repeated contact with the hormone[11]. The mechanism of 'fading receptor memory', or receptor desensitization, as it may also be called, is well known to occur in situations where a particular hormone has been removed from the circulation. The receptor remains sensitized only as long as it has repeated (sometimes pulsatile) contact with the hormone. In case of excess hormonal activity, membrane receptors may actually be 'down regulated', possibly as a result of endocytosis, thus avoiding hyperreaction of the cell.

Primordial binding sites may be located as proteins or glycoproteins within cellular membranes, or they may be dissolved in intracellular (cytoplasm) or extracellular (plasma) fluid. Receptor imprinting has not only been demonstrated in unicellular organisms, but also in mammalian cell lines and in multicellular organisms during development[9, 11]. Insulin induced imprinting of insulin receptors in mammalian cell lines at a very low concentration (10^{-13} M) and during the rather short exposure time of only one hour[14]. Similarly, TSH and FSH were able to imprint their receptors in Chinese hamster ovary cells in cell cultures. The authors[15] observed that the receptor imprinted by TSH was able to bind both TSH and FSH. The receptor which had been imprinted by FSH was also able to bind either hormone. This type of overlap in the imprinting effect was attributed to the structural similarity of the two glycoprotein hormones, which are known to share a common alpha subunit.

A great number of experiments have been performed on the imprinting effect of steroids, peptides, amino acids and cardioactive glycosides during pre- and postnatal development of rats and chickens[11]. From the results of these studies, the authors concluded that, even in higher organisms, imprinting of receptor systems is result of a first contact with the particular hormones. This imprinting, however, depends on particular sensitive phases during development of the organism and it depends on the concentration of the interacting hormone. The appropriate hormone can actually damage the receptor permanently if it appears at an unsuitable time and at an inappropriate concentration[11]. A similar conclusion was drawn by Döhler[18] in regard to sexual imprinting.

The observation that hormone-induced receptor imprinting is transmitted from one cell generation to the other, and was demonstrable in *Tetrahymena* after as many as 500 cell generations[11], indicates that hormonal imprinting of receptors is not only a mechanism limited to the cell membrane. Normal metabolic turnover of the membrane is so rapid that membrane-associated structures are replaced very rapidly. Csaba[10] concluded from his studies that receptors can be defined as genetically determined structures whose expression is subject to amplification by environmental influences. Amplification does not take place in isolated (hormone free) conditions.

Teleologically speaking, the synthesis of a hormone and its receptor within the same cell does not make sense. Hormones and receptors only interact during a short period of information exchange. For example, cell A releases hormone A, which carries a message for cell B. In order to understand the message, cell B has to develop a specific receptor for hormone A. If cell B wants to reply to cell A, then cell B has to release hormone B and cell A has to develop a receptor for hormone B. A very clear example for this type of interaction was given by

LeRoith et al.[42] in their description of mating factors in the budding yeast *Saccaromyces cerevisae*. This yeast exists in a diploid form as well as two haploid forms, designated 'alpha' and 'a'. Union of two haploids of opposite type is a process which requires two peptide pheromones or mating factors, 'alpha' factor and 'a' factor. 'Alpha' factor is produced by 'alpha' cells and acts only on cells of the opposite type ('a' cells), via specific cell surface receptors. 'A' factor is produced by 'a' cells and acts only on 'alpha' cells, via specific cell surface receptors[64]. Interestingly, 'alpha' factor bears a striking structural resemblance to mammalian gonodotropin releasing hormone (GnRH), binds to GnRH receptors of rat pituitary cells, and causes release of gonadotropins from the pituicytes[26, 47].

Another example, mentioned by LeRoith et al.[42], demonstrates the organizing effect of a hormone from one cell on the target organ of another cell. In the unicellular fungus *Achlya ambisexualis* the female secretes antheridiol, a steroid pheromone, which induces in the male the formation of antheridial branches (sex organs) and the secretion of oogoniol, a second pheromone. Oogoniol causes the female to develop sex organs[25].

A teleological point of view

The development of a receptor for a particular substance is necessary for a cell, teleologically speaking, only if uptake of this substance as an information carrier has a selective advantage for the cell. Vertebrate hormones are normally not present in the extracellular environment of unicellulars, and therefore there is no need for this organism to develop receptors for vertebrate hormones. If, however, the composition of the extracellular environment should change, then it may become advantageous for the organism to develop immediately specific receptors for the substances in the new environment. It would be very inefficient for the organism to wait for the remote chance of an accidentally occurring genetic mutation which, through the supply of a new receptor, would help the organism to adapt to the new environment. The predisposition of 'plastic' primordial receptors through genetic mechanisms and the final shaping (imprinting) of these receptors, depending on the particular environmental needs, would indeed be a 'fail-safe' mechanism through which nature provides optimum adaptability to changing environments. The same system would also work for multicellular organisms. During sex determination, for example, there is a genetic predisposition towards either male or female. For final development of organ structure and function, however, this genetic predisposition can be overruled by the priming action of circulating sex hormones[18].

During early development of an organism the shape of primordial membrane receptors may be oriented in line with neighboring structures of the cell membrane, thus forming a yet undifferentiated unit of optimal molecular energy distribution. One may envision a process in which the initial contact of a hormone with a primordial receptor is sufficient for recognition of the appropriate future binding site. After this recognition considerable induction of shape may take place, during which the electrostatic forces of the hormone and its hydropathic composition will force the receptor to take a shape providing

190

complementary electrostatic forces and complementary hydropathic form. The mapping of electrostatic potential onto molecular surfaces has revealed the importance of electrostatic forces in intermolecular interactions[65]. A similar mechanism has been suggested for the induction of antibodies by antigens[65].

1 Acevedo, M. F., Slifkin, M., Pouchet, G. R., and Pardo, M., Immunocytochemical localization of a choriogonadotropin like protein in bacteria isolated from cancer patients. Cancer *41* (1978) 1217–1219.
2 Barrington, E. J. W., Evolutionary aspects of hormonal structure and function, in: Comparative Endocrinology, pp. 381–396. Eds P. J. Gaillard and H. H. Boer. Elsevier/North Holland, Amsterdam 1978.
3 Barrington, E. J. W., The phylogeny of the endocrine system. Experientia *42* (1986) 775–782.
4 Berelowitz, M., LeRoith, D., Von Schenk, H., Newgard, C., Szabo, M., Frohman, L. A., Shiloach, J., and Roth, J., Somatostatin-like immunoactivity and biological activity is present in T. pyriformis, a ciliated protozoan. Endocrinology *110* (1982) 1939–1944.
5 Bhatnagar, Y. M., and Carraway, R., Bacterial peptides with C-terminal similarities to bovine neurotensin. Peptides *2* (1981) 51–59.
6 Blalock, J. E., and Smith, E. M., Hydropathic anti-complementarity of amino acids based on the genetic code. Biochem. biophys. Res. Commun. *121* (1984) 203–207.
7 Bost, K. L., Smith, E. M., and Blalock, J. E., Similarity between the corticotropin (ACTH) receptor and a peptide encoded by an RNA that is complementary to ACTH mRNA. Proc. natn. Acad. Sci. USA *82* (1985) 1372–1375.
8 Burshell, A., Stathis, P. A., Do, Y., Miller, S. C., and Feldman, D., Characterization of an estrogen-binding protein in the yeast Saccharomyces cervisiae. J. biol. Chem. *259* (1984) 3450–3456.
9 Csaba, G., Ontogeny and Phylogeny of Hormone Receptors. Monographs in Developmental Biology, vol. 15. Karger, Basel 1981.
10 Csaba, G., The present state in the phylogeny and ontogeny of hormone receptors. Horm. metab. Res. *16* (1984) 329–335.
11 Csaba, G., Receptor ontogeny and hormonal imprinting. Experientia *42* (1986) 750–759.
12 Csaba, G., Kovàcs, P., and Inczefi-Gonda, A., Insulin binding sites induced in the Tetrahymena by rat liver receptor antibody. Z. Naturf. *39c* (1984) 183–185.
13 Csaba, G., and Lantos, T., Effect of hormones on protozoa. Studies on the phagocytotic effect of histamin, 5-hydroxytryptamine and indoleacetic acid in Tetrahymena pyriformis. Cytobiologie *7* (1973) 361–365.
14 Csaba, G., Török, O., and Kovács, P., Hormonal imprinting in cell culture. I. Impact of single exposure to insulin on cellular insulin binding capacity in permanent cell lines. Acta physiol. hung. *64* (1984) 57–63.
15 Csaba, T., Török, O., and Kovács, P., Hormonal imprinting in cell culture. II. Induction of hormonal imprinting and thyrotropin (TSH) – gonadotropin (FSH) overlap in a Chinese hamster ovary (CHO) cell line. Acta physiol. hung. *64* (1984) 135-138.
16 Deftos, L., LeRoith, D., Shiloach, J., and Roth, J., Salmon, Calcitonin-like immunoactivity in extracts of *Tetrahymena pyriformis*. Horm. Metab. Res. *17* (1985) 82–85.
17 Dix, C. J., and Jordan, V. C., Modulation of rat uterine steroid hormone receptors by estrogen and antiestrogen. Endocrinology *107* (1980) 2011–2020.
18 Döhler, K. D., The special case of hormonal imprinting, the neonatal influence of sex. Experientia *42* (1986) 759–769.
19 Duve, H., Thorpe, A., and Lazarus, N. R., Isolation of material displaying insulin-like immunological and biological activity from the brain of the blowfly, *Calliphora vomitoria*. Biochem. J. *184* (1979) 221–227.
20 Frowein, J., Engel, W., and Weise, H. C., HCG receptor present in the gonadotropin insensitive Leydig cell of the immature rat. Nature, new Biol. *246* (1973) 141–150.
21 Fukushima, J., Watanabe, S., and Kaushima, K., Extraction and purification of substance with luteinizing hormone releasing activity from the leaves of avenasativa. Tohoku J. exp. Med. *119* (1976) 115–122.
22 Gorbman, A., and Bern, H. A., A Textbook of Comparative Endocrinology, Wiley & Sons, New York 1962.

23 Gorski, J., The nature and development of steroid hormone receptors. Experientia *42* (1986) 744–750.

24 Gorski, J., and Gannon, F., Current models of steroid hormone action: A critique. A. Rev. Physiol. *38* (1976) 425–450.

25 Horgan, P. A., The role of the steroid sex pheromone antheridiol in controlling the development of male sex organs in the water mold, Achlya, in: Sexual Interactions in Eukaryotic Microbes, Eds D. H. O'Day and P. A. Horgen, pp. 155–178. Academic Press, New York 1981.

26 Hunt, L. T., and Dayoff, M. D., Structural and functional similarities among hormones and active peptides from distantly related eukaryotes, in: Peptides: Structure and Biological Function, Proceedings of the Sixth American Peptide Symposium, Eds E. Gross and J. Meienhofer, pp. 757–760. Pierce Chemical Co, Rockford, Illinois 1979.

27 Jackson, I. M. D., Abundance of immunoreactive thyrotropin-releasing hormone-like material in the alfalfa plant. Endocrinology *108* (1981) 344–346.

28 Jordan, V. C., Dix, C. J., Rowsby, L., and Prestwich, G., Studies on the mechanism of action of the nonsteroidal antiestrogen tamoxifen in the rat. Molec. cell. Endocr. *7* (1977) 177–192.

29 Jordan, V. C., Prestwich, G., Dix, C. J., and Clark, E. R., Binding of anti-estrogens to the estrogen receptor, the first step in anti-estrogen action, in: Pharmacological Modulation of Steroid Action, pp. 81–98. Eds E. Genazzani, F. DiCarlo and W. I. P. Mainwaring. Raven Press, New York 1980.

30 Josefsson, J. O., and Johansson, P., Naloxone-reversible effects of opioids on pinocytosis in amoebae proteus. Nature *282* (1979) 78–80.

31 Kaiser, E. T., and Kezdy, F. J., Amphiphilic secondary structure: design of peptide hormones. Science *223* (1984) 249–255.

32 Karin, M., Haslinger, A., Holtgreve, H., Richards, R. I., Krauter, P., Westphal, H. M., and Beato, M., Characterization of DNA sequences through which cadmium and glucocorticoid hormones induce human metallothionein-IIA gene. Nature *308* (1984) 513–519.

33 Kassis, J. A., Sakai, D., Walent, J. H., and Gorski, J., Primary cultures of estrogen-responsive cells from rat uteri: Induction of progesterone receptors and a secreted protein. Endocrinology *114* (1984) 1558–1566.

34 Kato, J., Onouchi, T., and Okinaga, S., Hypothalamic and hypophysial progesterone receptors: Estrogen-priming effect, differential localization, 5 alpha-dihydroprogesterone binding, and nuclear receptors. J. Steroid Biochem. *9* (1978) 419–427.

35 Kovács, P., The mechanism of receptor development as implied from hormonal imprinting studies on unicellulars. Experientia *42* (1986) 770–775.

36 Kovács, P., and Csaba, G., Receptor level study of polypeptide hormone (insulin, TSH, FSH) imprinting and overlap in *Tetrahymena*. Acta protozool., in press (1985).

37 Kovács, P., Csaba, G., and Bohdaneczky, E., Immunological evidence of the induced insulin receptor in *Tetrahymena*. Comp. Biochem. Physiol. *80A* (1985) 41–42.

38 LeRoith, D., Lesniak, M. A., and Roth, J., Insulin in insects and annelids. Diabetes *30* (1981) 70–76.

39 LeRoith, D., Berelowitz, M., Pickens, W., Crosby, L. H., and Shiloach, J., Somatostatin-related material in E. coli: Evidence for two molecular forms. Biochim. biophys. Acta, in press (1985).

40 LeRoith, D., Liotta, A. S., Roth, J., Shiloach, J., Lewis, M. E., Pert, C. B., and Krieger, D. T., Corticotropin and *β* endorphin-like materials are native to unicellular organisms. Proc. natn. Acad. Sci. USA *79* (1982) 2086–2090.

41 LeRoith, D., Pickens, W., Wilson, G. L., Miller, B., Berelowitz, M., Vinik, A. I., Collier, E., and Cleland, C. F., Somatostatin-like material is native to flowering plants. Endocrinology *117* (1985) 2093–2097.

42 LeRoith, D., Roberts, C. Jr, Lesniak, M. A., and Roth, J., Receptors for intercellular messenger molecules in microbes: Similarities to vertebrate receptors and possible implications for diseases in man. Experientia *42* (1986) 782–788.

43 LeRoith, D., Shiloach, J., Roth, J., and Lesniak, M. A., Evolutionary origins of vertebrate hormones: Substances similar to mammalian insulins are native to unicellular organisms. Proc. natn. Acad. Sci. USA *77* (1980) 6184–6188.

44 LeRoith, D., Shiloach, J., Roth, J., and Lesniak, M. A., Insulin or a closely related molecule is native to *Escherichia coli*. J. biol. Chem. *256* (1981) 6533–6536.

45 Loose, D. S., and Feldman, D., Characterization of a unique corticosterone-binding protein in Candida albicans. J. biol. Chem. *257* (1982) 4925–4930.

46 Loose, D. S., Stover, E. P., Restrepo, A., Stevens, D. A., and Feldman, D., Estradiol binds to a receptor-like cytosol binding protein and initiates a biological response in Paracoccidioides brasiliensis. Proc. natn. Acad. Sci. USA *80* (1983) 7659–7663.

47 Loumaye, E., Thorner, J., and Catt, K. J., Yeast mating pheromone activates mammalian gonad-otrophs: Evolutionary conservation of a reproductive hormone. Science *218* (1982) 1324–1325.

48 Martin, G., Keller, R., Kegel, G., Besse, G., and Jaros, P. P., The hyperglycemic neuropeptide of the terrestrial isopod Porcellio dilatatus. I. Isolation and characterization. Gen. comp. Endocr. *55* (1984) 208–216.

49 Maruo, T., Cohen, H., Segal, S. J., and Koide, S. S., Production of choriogonadotropin-like factor by a microorganism. Proc. natn. Acad. Sci. USA *76* (1979) 6622–6626.

50 Mordue, W., and Stone, J. N., Insect hormones in: Hormones and Evolution. Ed. E. J. W. Barrington, pp. 215–271. Academic Press, London 1979.

51 Morley, J. E., Meyer, N., Pekary, A. E., Melmed, S., Carlso, H. E., Briggs, J. E., and Hershman, J. M., A prolactin inhibitory factor with immuno-characteristics similar to thyrotropin releasing factor (TRH) is present in rat pituitary tumors (GH3 & W5), testicular tissue, and a plant material, Alfalfa. Biochem. biophys. Res. Commun. *96* (1980) 47–53.

52 Muggeo, M., Ginsberg, B. H., Roth, J., Neville, D. M., De Meyts, P., and Kohn, C. R., The insulin receptor in vertebrates is functionally more conserved during evolution than the insulin itself. Endocrinology *104* (1979) 1393–1402.

53 Mulvihil, E. R., LePennec, J.-P., and Chambon, P., Chicken oviduct progesterone receptor: location of specific regions of high-affinity binding in cloned DNA fragments of hormone-respon-sive genes. Cell *28* (1982) 621–632.

54 Payvar, F., DeFranco, D. G., Firestone, L., Edgar, B., Wrange, D., Okret, S., Gustafsson, J.-A., and Yamamoto, K. R., Sequence-specific binding of glucocorticoid receptor to MTV DNA at sites within and upstream of the transcribed region. Cell *35* (1983) 381–392.

55 Perez-Cano, R., Murphy, P. K., Girgis, S. I., Arnett, T. R., Blankharn, I., and MacIntyre, I., Unicellular organisms contain a molecule resembling human calcitonin. Endocrinology *110* (1982) 673 (abstract).

56 Petruzzelli, L., Herrera, R., Garcia, R., and Rosen, O. M., The insulin receptor of *Drosophila melanogaster*. Cold Spring Harbor Symp., in press (1985).

57 Plisetskaya, E., Kazakov, V. K., Solititskaya, L., and Leibson, L. G., Insulin producing cells in the gut of freshwater bivalve molluscs *Anodonta cygnea* and *Unio pictorum* and the role of insulin in the regulation of their carbohydrate metabolism. Gen. comp. Endocr. *35* (1978) 133–145.

58 Posner, B. I., Kelly, P. A., and Friesen, H. G., Induction of a lactogenic receptor in rat liver: influence of estrogen and the pituitary. Proc. natn. Acad. Sci. USA *71* (1974) 2407–2410.

59 Richert, N. D., and Ryan, R. J., Specific gonadotropin binding to Pseudomonas maltophilia. Proc. natn. Acad. Sci. USA *73* (1977) 878–882.

60 Rubinovitz, C., and Shiloach, J., Insulin related material in prokaryotes. FEMS Microbiol. Lett., in press (1985).

61 Scarborough, R. M., Jamieson, G. C., Kalish, F., Kramer, S. J., McEnroe, G. A., Miler, C. A., and Schooley, D. A., Isolation and primary structure of two peptides with cardioacceleratory and hyperglycaemic activity from the corpora cardiaca of Periplaneta americana. Proc. natn. Acad. Sci. USA *81* (1984) 5575–5579.

62 Schwabe, C., LeRoith, D., Thompson, R. P., Shiloach, J., and Roth, J., Relaxin extracted from protozoa (Tetrahymena pyriformis). J. biol. Chem. *258* (1983) 2778–2782.

63 Sela, I., Plant-virus interaction related to resistance and localization of viral infections. Adv. Virus Res. *26* (1981) 201–237.

64 Stotzler, D., and Duntze, W., Isolation and characterization of four related peptides exhibiting alpha factor activity from Saccharomyces cerevisaiae. Eur. J. Biochem. *65* (1976) 257–262.

65 Tainer, J. A., Getzoff, E. D., Alexander, H., Houghten, R. A., Olson, A. J., Lerner, R. A., and Hendrickson, W. A., The reactivity of anti-peptide antibodies is a function of the atomic mobility of sites in a protein. Nature *312* (1984) 127–134.

66 Van Noorden, S., and Polak, J. M., Hormones of the alimentary tract, in: Hormones and Evolution 2, pp. 791–828. Ed. E. J. W. Barrington. Academic Press, London 1979.

67 Wallis, M., The molecular evolution of pituitary hormones. Biol. Rev. *50* (1975) 35–98.

68 Weiss, M., Ingbar, S. H., Winblad, S., and Kasper, D. L., Demonstration of a saturable binding site for thyrotropin in Yersinia enterocolitica. Science *219* (1983) 1331–1333.

69 Yamamoto, K. R., and Alberts, B. M., Steroid receptors: Elements for modulation of eukaryotic transcription. A. Rev. Biochem. *45* (1976) 721–746.

70 Zioudrov, C., Streaty, R. A., and Klee, W. A., Opioid peptides derived from food proteins. The exorphins. J. biol. Chem. *254* (1979) 2446–2450.

Index

DATE DUE

FEB 1 1991		
JUN 0 2 1992		
JUN 8 1993		
JUL 1 4 2022		
SEP 0 7 1993		